T0335936

Information Security and Cryptography

More information about this series at http://www.springer.com/series/4752

Joel Reardon

Secure Data Deletion

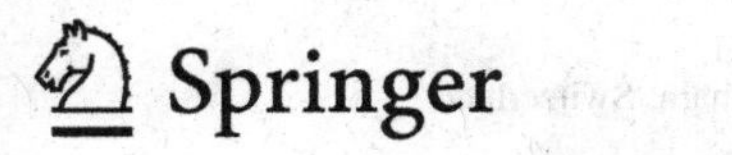

Joel Reardon
Institute of Information Security
ETH Zürich
Zürich
Switzerland

ISSN 1619-7100 ISSN 2197-845X (electronic)
Information Security and Cryptography
ISBN 978-3-319-28777-5 ISBN 978-3-319-28778-2 (eBook)
DOI 10.1007/978-3-319-28778-2

Library of Congress Control Number: 2016955922

Printed on acid-free paper

This Springer imprint is published by Springer Nature
The registered company is Springer International Publishing AG
The registered company address is: Gewerbestrasse 11, 6330 Cham, Switzerland

To the wonderful teachers I've had.

Acknowledgments

I would like to acknowledge the contributions of many people to the conception and completion of this book. Foremost is my doctoral advisors Srđan Čapkun and David Basin, with whom the contents of this book were conceived, conjectured and criticized, along with additional collaboration with Claudio Marforio, Alina Oprea, and Hubert Ritzdorf. Various drafts of chapters and the papers on which they are based received helpful comments from Steven Dropsho, Clemens Fruhwirth, Arthur Gervais, Marco Guarnieri, Ari Juels, Nikolaos Karapanos, Gregor Kasieczka, Kari Kostiainen, René Kyllingstad, Luka Malisa, Grgur Maretic, Ognjen Maric, Srdjan Marinovic, Ramya Jayaram Masti, Christina Pöpper, Aanjhan Ranganathan, Claudio Soriente, Alessandro Sorniotti, Nils Ole Tippenhauer, Thomas Themel, Petar Tsankov, Paul Van Oorschot, Thilo Weghorn, and Der-Yeuan Yu. Abel Groenewolt improved the aethetics of the figures by commenting on their clarity, consistency, and symmetry. Ronan Nugent provided great support from Springer in producing this work. I gratefully acknowledge the Zurich Information Security Center and the Swiss National Science Foundation for their support of this work. Finally, I would like to give my deepest gratitude to Joanna Bolechowska as well as my family for their love and support.

Contents

Part I Introduction and Background

1 Introduction 3
1.1 Organization and Structure 7

2 Related Work on Secure Deletion 11
2.1 Introduction 11
2.2 Related Work 11
2.2.1 Layers and Interfaces 12
2.2.2 Physical-Layer and Controller-Layer Sanitization 15
2.2.3 User-Level Solutions 17
2.2.4 File-System-Level Solutions with In-Place Updates 20
2.2.5 Cross-layer Solutions 22
2.2.6 Summary 23
2.3 Adversarial Model 23
2.3.1 Classes of Adversarial Capabilities 23
2.3.2 Summary 25
2.4 Analysis of Solutions 26
2.4.1 Classes of Environmental Assumptions 27
2.4.2 Classes of Behavioural Properties 27
2.4.3 Summary 30

3 System Model and Security Goal 33
3.1 Introduction 33
3.2 System Model 33
3.3 Storage Medium Models 34
3.4 Adversarial Model 36
3.5 Security Goal 37

Part II Secure Deletion for Mobile Storage

4 Flash Memory: Background and Related Work ... 47
4.1 Overview ... 47
4.2 Flash Memory ... 48
4.2.1 In-Place Updates and Log-Structured File Systems ... 49
4.2.2 Flash Translation Layer ... 51
4.2.3 Flash File Systems ... 52
4.2.4 Generalizations to Other Media ... 52
4.3 Related Work for Flash Secure Deletion ... 52
4.4 Summary ... 55

5 User-Level Secure Deletion on Log-Structured File Systems ... 57
5.1 Introduction ... 57
5.2 System and Adversarial Model ... 58
5.3 YAFFS ... 58
5.4 Data Deletion in Existing Log-Structured File Systems ... 59
5.4.1 Instrumented YAFFS ... 60
5.4.2 Simulating Larger Storage Media ... 61
5.5 User-Space Secure Deletion ... 63
5.5.1 Purging ... 64
5.5.2 Ballooning ... 65
5.5.3 Hybrid Solution: Ballooning with Purging ... 67
5.6 Experimental Evaluation ... 67
5.6.1 Experimental Results ... 68
5.7 Summary ... 72
5.8 Research Questions ... 72

6 Data Node Encrypted File System ... 73
6.1 Introduction ... 73
6.2 System and Adversarial Model ... 73
6.3 DNEFS's Design ... 74
6.3.1 Key Storage Area ... 75
6.3.2 Keystore ... 75
6.3.3 Clocked Keystore Implementation ... 77
6.3.4 Clock Operation: KSA Update ... 79
6.3.5 Key-State Map ... 79
6.3.6 Summary ... 81
6.4 Extensions and Optimizations ... 81
6.4.1 Granularity Trade-off ... 81
6.4.2 KSA Update Policies ... 83
6.4.3 KSA Organization ... 83
6.4.4 Improving Reliability ... 83
6.4.5 Encrypted File System ... 84
6.5 Summary ... 84

6.6 Research Questions 85

7 UBIFSec: Adding DNEFS to UBIFS 87
7.1 Introduction 87
7.2 System and Adversarial Model 87
7.3 Background 87
7.3.1 MTD and UBI Layers 88
7.3.2 UBIFS 88
7.4 UBIFSec Design 90
7.4.1 Key Storage Area 90
7.4.2 Key-State Map 92
7.4.3 Summary 94
7.5 Experimental Validation 94
7.5.1 Android Implementation 96
7.5.2 Wear Analysis 96
7.5.3 Power Consumption 98
7.5.4 Throughput Analysis 98
7.5.5 Timing Analysis 99
7.6 Conclusions 101
7.7 Practitioner's Notes 101

Part III Secure Deletion for Remote Storage

8 Cloud Storage: Background and Related Work 105
8.1 Introduction 105
8.2 Persistent Storage 105
8.2.1 Securely Deleting and Persistent Combination 106
8.2.2 Cloud Storage 106
8.3 Related Work 108
8.4 Summary 113

9 Secure Data Deletion from Persistent Media 115
9.1 Introduction 115
9.2 System and Adversarial Model 116
9.3 Graph Theory Background 116
9.4 Graph-Theoretic Model of Key Disclosure 118
9.4.1 Key Disclosure Graph 118
9.4.2 Secure Deletion 120
9.5 Shadowing Graph Mutations 120
9.5.1 Mangrove Preservation 123
9.5.2 Shadowing Graph Mutation Chains 125
9.5.3 Mangrove Key Disclosure Graphs in Related Work 126
9.6 Summary 127
9.7 Research Questions 128

10 B-Tree-Based Secure Deletion 129
10.1 Introduction 129
10.2 System and Adversarial Model 130
10.3 Background 130
10.3.1 B-Tree Storage Operations 130
10.3.2 B-Tree Balance Operations 131
10.4 Securely Deleting B-Tree Design 131
10.4.1 Cryptographic Details 132
10.4.2 Data Integrity 132
10.4.3 Versioning 133
10.4.4 Skeleton Tree 133
10.4.5 Commitment 135
10.4.6 Crash Safety 135
10.5 Implementation Details 136
10.5.1 Data Storage 136
10.5.2 Network Block Device 137
10.5.3 Virtual Storage Device 137
10.5.4 Caches 137
10.6 Experimental Evaluation 138
10.6.1 Workloads 138
10.6.2 Caching 139
10.6.3 B-Tree Properties 140
10.7 Conclusions 140
10.8 Practitioner's Notes 141

11 Robust Key Management for Secure Data Deletion 143
11.1 Introduction 143
11.2 System and Adversarial Model 144
11.2.1 System Entities 145
11.2.2 Adversarial Model 146
11.3 Distributed Keystore 147
11.3.1 Distributed Clocked Keystore 147
11.3.2 Distributed Keystore Correctness 148
11.4 Synchronization 150
11.5 Byzantine Robustness 153
11.6 Keystore Secure Deletion 156
11.6.1 Key Pools and Encryption Keys 157
11.6.2 Encryption Key Encoding 159
11.6.3 XOR-Based Encoding 159
11.6.4 Security Analysis 162
11.7 Implementation Details 166
11.8 Experimental Validation 170
11.9 Conclusions 173
11.10 Research Questions 173
11.11 Practitioner's Notes 174

Part IV Conclusions

12 Conclusion and Future Work 177
12.1 Summary of Contributions 177
12.2 Related and Complementary Research 177
12.2.1 Information Deletion 179
12.2.2 File Carving 180
12.2.3 Steganographic and Deniable Storage 180
12.2.4 History Independence 181
12.2.5 Provable Deletion 182
12.3 Future Work 182
12.3.1 New Types of Storage Media 182
12.3.2 Benchmarks for Different Storage 183
12.3.3 Secure-Deletion Data Structure Selection 183
12.3.4 Formalization 184
12.3.5 DNEFS for FTLs 184
12.4 Concluding Remarks 185

References 187

Glossary 193

Index 201

Acronyms

AES	Advanced Encryption Standard; a widely supported block cipher
AT	access token; a secret binary string that grants access to some data
CA	certificate authority; in public-key cryptography, a highly trusted entity that checks the validity of certificates and issues cryptographic signatures attesting to their validity
CTR	counter; a block cipher mode that turns the cipher into a stream cipher by using the encryption key to encrypt a sequence of numbers, the ciphertext of which is XORed to the plaintext
DB	database; an organized collection of data and relations
DEFY	deniable encrypted file system from YAFFS; a deniable file system that uses secure deletion
DHT	distributed hash table; a form of key-value storage consisting of a collection of nodes, each of which partially stores the key space
DNEFS	data node encrypted file system; a file system we propose that encrypts each unit of data with its own unique encryption key
DoS	denial of service; an attack on the availability of a system
DVD	digital versatile disc; an optical storage medium that, in its common form, cannot be modified once written
ECB	electronic codebook; a block cipher mode where each plaintext block maps to the same ciphertext block
F2FS	flash-friendly file system; a Linux file system designed for flash memory
FAT	file allocation table; a simple and widely supported file system
FFS	flash file system; a file system designed specifically for the nuances of flash memory in the file system
FTL	flash translation layer; a block-level remapping for flash memory that allows non-flash file systems to use the memory
FUSE	file systems in user space; a Linux feature that allows a user-space program to execute as a mountable file system
HTTP	hypertext transfer protocol; a popular Internet protocol used for the World Wide Web
HTTPS	HTTP secure; the integration of TLS with HTTP

I/O	input/output; the flow of data through a computer system
IOCTL	IO control; an OS function that permits low-level device-specific operations that are not available in a standard interface
IV	initialization vector; a non-secret random string used in some encryption functions
JFFS	journalling flash file system; a file system designed for flash memory that is part of the mainline Linux kernel
KDG	key disclosure graph; a graph we developed to reason about the worst-case adversarial knowledge in a system that uses encryption and key wrapping
KP	key position; a reference to a location where a key value is stored
KSA	key storage area; an area of memory reserved for storing encryption keys
KSFS	keystore file system; a file system we developed that uses a keystore for secure deletion
KV	key value; a binary string suitable for use as an encryption key
LEB	logical erase block; in UBI, the interface's erase blocks that neither suffer wear nor can be destroyed
LFU	least-frequently used; a caching strategy that evicts the least frequently used piece of data
LRU	least-recently used; a caching strategy that evicts the least recently used piece of data
LZO	Lempel–Ziv–Oberhumer; a common data compression algorithm
MMC	multimedia card; the generic name for flash-memory-based removable storage cards
MTD	memory technology device; the term used to refer to flash memory storage in Linux
NDB	network block device; a virtual block device that forwards all I/O requests over a network socket
OSD	object store device; a storage system that organizes data as objects that include their own metadata
PEB	physical erase block; in UBI, the actual flash erase blocks that are logically remapped by UBI
POSIX	portable operating system interface; an abstract operating system and core set of operating system features used for portability across implementations
RPC	remote procedure call; a software technique to execute a function on a remote computer
SCSI	small computer system interface; a standard interface for communication between a computer and peripheral hardware such as disk drives
SSD	solid-state drive; a storage medium technology that uses flash memory and replaces hard drives
TCP	transmission control protocol; a network communication protocol that provides reliable in-order delivery of a stream of data
TLS	transport layer security; a set of cryptographic protocols used to provide secrecy and authenticity over computer networks

TNC	tree node cache; in UBIFS, an in-memory data structure used to speed up file system operations
UBI	unordered block images; an interface for MTD devices that adds a logical remapping and other features
UBIFS	UBI file system; a file system designed for the UBI interface
VM	virtual machine; an emulation of computer hardware done in software
WORM	write-once, read-many; a type of storage media that cannot be deleted or updated
XOR	exclusive OR; a boolean operation on two boolean operands that returns true if and only if exactly one of the operands is true
YAFFS	yet another flash file system; a file system designed for flash memory and used on some Android devices

Part I
Introduction and Background

Chapter 1
Introduction

During New York City's 2012 Thanksgiving Day parade, sensitive personal data rained from the sky. Makeshift confetti, formed out of shredded police case reports and personnel files, landed on spectators who observed something peculiar about it: having been shredded horizontally, entire stretches of text (names, social security numbers, arrest records, etc.) were completely legible [1]. It is likely that the documents were shredded to *securely delete* the sensitive data they contained (and not simply to make confetti).

Secure data deletion is the task of deleting data from a storage medium (i.e., anything that stores data, such as a hard drive, a phone, a brain, or a blackboard) so that the data is *irrecoverable*. This irrecoverability is what distinguishes secure deletion from other forms of deletion, which may, for example, delete data to reclaim storage resources. We *securely* delete data with a security goal in mind: to prevent our adversary from gaining access to it. As is common in security, the steps and precautions we take increase with the sophistication of our adversary and the gravity of the consequences of our data's exposure. Giving an old computer to a family member may warrant no effort while discarding hard drives containing medical data protected by strict regulation may warrant a great deal more care. Secure data deletion is therefore a natural and necessary requirement in the storage of confidential data. This book examines in detail this seemingly simple problem and shows that many complicated issues lie under the surface.

The problem of secure deletion is known in the scientific literature, which has given it many names. Thus we hear of data being assuredly deleted [2–4], completely removed [3, 5], deleted [6], destroyed [2, 5, 7], erased [5–8], expunged [9], forgotten [2, 5], permanently erased [10], purged [4, 11], reliably removed [7], revoked [3, 5], sanitized [7, 11], securely erased [12], self-destructed [13, 14], and, of course, securely deleted [15, 16]. Whether explicitly stated as a system requirement or implicitly assumed, and however named, an ability to securely delete data in the presence of an adversary is required for the security of many systems.

J. Reardon, *Secure Data Deletion*, Information Security and Cryptography,
DOI 10.1007/978-3-319-28778-2_1

The Need to Delete.

In the physical world, the importance of secure deletion is well understood: paper shredders are single-purpose devices widely deployed in offices; published government information is selectively redacted; access to top-secret documents is managed to ensure that all copies can be destroyed when necessary. In the digital world, the importance of secure deletion is also well recognized. Legislative or corporate requirements can require secure deletion of data prior to disposing of or selling hard drives. This is particularly so when the data is considered to be sensitive, for example, health data, financial data, trade secrets, or privileged communications.

To achieve secure deletion correctly, the National Institute of Standards and Technology (NIST) has standardized secure deletion best practices across a wide variety of storage media [11]. NIST's techniques are designed to securely delete the entire storage medium, and some techniques result in the medium's physical destruction: incineration, pulverization, etc. Destruction is an acceptable method if the storage medium does not need to be reused and the owner is provided sufficient time and warning to perform the secure deletion before surrendering the storage medium (or its leftover pieces).

NIST's techniques, however, are not a panacea as secure deletion is not limited to one-off events involving the secure deletion of all stored data. It is instead rather common that secure deletion is needed at a fine granularity and on a continuous basis, while the main system continues functioning normally—that is, not all of the system's data was deleted. We illustrate this with the following examples.

Mobile phones and other portable devices store a wide range of sensitive information, including the timestamped sequence of nearby wireless networks that effectively encodes the user's location history. In the popular web browsers Firefox [17], Chrome [18], and Safari [19], however, clearing the web browsing history is categorized as a *privacy option*. Mobile phones can delete individual text messages, clear call logs, delete photographs, etc. Corporate e-mails include a boilerplate footnote demanding the immediate deletion of the message if sent to an unintended recipient. This may be challenging if the mobile phone has already synchronized the email and written it to a storage medium incapable of providing secure deletion guarantees.

Network service operators collect logs for intrusion detection or other administrative purposes. However, a privacy-focused network service (such as an anonymous message board, mix network, or Tor relay) should wish to promptly securely delete log data once it has served its purpose; this necessitates partial secure deletion on a continuous basis. Network services may also need secure deletion simply to comply with regulations regarding their users' private data. Two examples are the European Union's *right to be forgotten* [20] that would force companies to store personal data in a manner that supports the secure deletion of a particular user's data upon request, and California's legislation that enforces similar requirements only for minors. Such regulations on private data may change, or new ones become enforced, turning data assets into data liabilities. This results in an immediate need to securely delete a significant amount of data, exemplified by the United Kingdom's demand that Google securely delete Wi-Fi data illegally collected by Google's Street View cars, wherever and however it was stored [21]. We will see that this is harder to achieve after the fact if the manner of the data's storage was not designed to facilitate on-demand fine-grained secure data deletion.

Fine-grain secure deletion is also needed as a basic operation to achieve other security properties, which we illustrate with two examples: *forward secrecy* and *proactive security*. Forward secrecy is the desirable property that ensures that the compromise of a user's long-term cryptographic key does not affect the confidentiality of past communications. This is often achieved by protecting the communications with session keys authentically negotiated using the long-term key. Proactive security [22] is the desirable property for long-term secret sharing systems that makes it harder for an adversary to compromise sufficient shares to determine the secret. This is done by periodically refreshing the shares such that old shares cannot be meaningfully combined with new shares; the adversary must therefore compromise all necessary shares within the time between refreshes—much harder than within the lifetime of the secret itself. In both of these cases, secure deletion is assumed to be a storage feature—a feature that is in fact critical to the system's correct operation.

Challenges.

As is often the case in the digital world, a straightforward problem is fraught with challenges and complications, and secure deletion is no exception. Digital data is effortlessly replicated, often without any record. Simply *finding* where data is stored over a vast number of computer systems and storage media may present a logistical nightmare, particularly when servers replicate data, go offline indefinitely, crash during copy operations, and have their hardware swapped around. Even a single copy on a single hard drive may be duplicated without notice, for instance, when the file system rearranges its storage during defragmentation.

Even when all the locations where data is stored can be found, it may not be possible to securely delete the data. Overwriting magnetic data may leave analog remnants accessible to adversaries with forensic equipment. Flash memory cannot efficiently be overwritten, and so new versions of files are instead written to a new location with the old one left behind. High-capacity magnetic tapes must be written end-to-end; worse, they are often then shipped off to a vault for off-line archiving. Optical discs like DVDs are a kind of WORM medium (a "write-once, read-many" medium) and such media only achieve secure deletion through physical destruction. The steps to achieve secure deletion therefore vary depending on the actual storage medium being used.

Another challenge in secure deletion is that many users are unaware that additional steps are needed to sanitize their storage media. Secure deletion seems simple: all modern file systems allow users to "delete" their files. Users of these systems may reasonably, but falsely, assume that when they delete the data, it is thenceforth irrecoverable [23]. Instead, "deletion" is a warning and not a promise. File systems implement deletions by *unlinking* files and *discarding* data blocks. Abstractly, unlinking a file only changes file system metadata to state that the file is now "deleted"; the file's full contents remain available. This is done for efficiency reasons—deleting a file would require changing all its data, while unlinking a file can be implemented by changing even only one bit (e.g., YAFFS1's "not deleted" bit). Since the beginning, file system designers have consistently made the assumption that the only reason that a user may want to delete a file is to recover the wasted storage resources so they may be allocated to new files. (Yet clearing the web cache is a *privacy* option!) Unlinking results in storage resources that are assumed to be free, but it is only when these resources are actually needed to store new data that that the discarded data they contain is finally securely deleted.

Even for those who know the best practices for secure deletion, the nature of digital information makes it hard to verify that the data is indeed irrecoverable. The user interfaces for deleting digital data simply do not provide the same rapid assurance of secure deletion as does a pile of (vertically) shredded mail. Forensic investigators of Chelsea Manning's laptop, for instance, discovered that she had tried to securely delete the contents of her laptop by overwriting its contents 35 times—an aggressive approach—but, unknown to her, the operation had stopped midway and left most of the data intact [24].

Garfinkel and Shelat [25] study secure deletion in practice. They include a forensic analysis of 158 used hard drives bought on the secondary market from 2000–2002. They discovered at that time that even the most basic sanitization is rarely employed. The kinds of data they easily found includes client documents from a law firm, a database of mental health patients, draft manuscripts, financial transactions executed by a bank machine, and plenty of credit card numbers. With little cost and effort, they were able to recover an extraordinary amount of personal information.

Garfinkel and Shelat speculate a variety of reasons why this may be the case: a lack of knowledge, of training, of concern for the problem or the data, or of tools to perform secure deletion. Although 52 of their hard drives were freshly formatted, it is unclear whether this was done as an attempt at secure deletion. The authors note that in "many interviews, users said that they believed DOS and Windows format commands would properly remove all hard drive data." Formatting warns that all data becomes irrecoverable, but in reality it only writes to a negligible fraction of the storage medium. In our work we aim to provide secure deletion solutions that allow users to easily and efficiently securely delete their data on a continuous basis.

Scope.

Abstractly, the user stores and operates on data items on a storage medium through an interface. *Data items* are addressable units of data; these include data blocks, database records, text messages, file metadata, entire files, entire archives, etc. The *storage medium* is any device capable of storing and retrieving these data items, such as a magnetic hard drive, a USB stick, or a piece of paper. The *interface* is how the user interacts with the storage medium; the interface offers functions to transform the user's data objects into a form suitable for storage on the storage medium. This transformation can also include operations such as encryption, error-correction, replication, etc. Our work focuses on how such data items are stored on storage media and how to modify the interface and implementations to effect secure data deletion.

This book presents a detailed treatment of secure deletion from years of research. We focus on secure deletion in the context of two types of storage media: mobile (e.g., smartphone) storage and remote (e.g., cloud) storage. These storage types represent a significant shift in how people access their data. Both lack secure data deletion despite being environments where users store their sensitive data and where adversaries can get access. Despite the massive difference in scale between the smartphone and the cloud, the problems and challenges of secure deletion share surprising similarities. Our goal is building efficient secure deletion solutions for these settings. In addition to our results, we list unanswered questions and notes to practitioners who may integrate these results into deployed systems.

Our scope is characterized by the following assumption: the contents of data items are independent of where it, and other items, are or were stored, and independent of what is or was stored at other storage positions. As an example, we assume that redacting text from a document is effective at securely deleting the redacted text, and our focus is on developing methods to perform this redaction. This is not always true: file systems store copies of data as temporary files, and functions computed over data content can unexpectedly appear as metadata. The redacted document may be filled in by either context or copies. Despite this, secure data deletion as a primitive operation is—at minimum—required to actually delete copies and mutations of data or metadata. Our scope is this secure data deletion.

1.1 Organization and Structure

This book is organized into four parts and 12 chapters.

Chapter 2 surveys related work and organizes existing solutions in terms of their interfaces. The chapter further presents a taxonomy of adversaries differing in their capabilities as well as a systematization of the characteristics of secure deletion solutions. Characteristics include environmental assumptions and behavioural properties of the solution.

Chapter 3 builds a system and adversarial model based on the survey of related work. This is the model that we use throughout this book. It also presents different types of storage media and illustrates the adversary's abilities and the user's goal.

Chapter 4 opens the part on secure deletion for mobile storage. It first presents details on the characteristics of flash memory, which is currently ubiquitously used in portable storage devices. Flash memory has the problem that the unit of erasure is much larger than the unit of read and write, and worse, erasure is expensive. It then presents related work for flash memory as well as generalizations of this erasure asymmetry to other kinds of media.

Chapter 5 presents our research into user-level secure deletion for flash memory, with a concrete example of an Android-based mobile phone. We show that these systems provide no timely data deletion, and that the time data remains increases with the storage medium's size. We propose two user-level solutions that achieve secure deletion as well as a hybrid of them, which guarantees periodic, prompt secure data deletion regardless of the storage medium's size. We also develop a model of the writing behaviour on a mobile device that we use to quantify our solution's performance.

Chapter 6 presents DNEFS, a file system change that provides fine-grained secure data deletion and is particularly suited to flash memory. DNEFS encrypts each individual data item and colocates all the encryption keys in a densely packed key storage area. DNEFS is efficient in flash memory erasures because the expensive erasure operation is only needed for the key storage area.

Chapter 7 presents UBIFSec, an implementation of DNEFS with the flash file system UBIFS. We describe our implementation and furthermore integrate UBIFSec in the Android operating system. We measure its performance and show that it is a usable and efficient solution. Android OS and applications run normally when using UBIFSec as the file system.

Chapter 8 begins the part on secure deletion for remote storage. We present details on the characteristics of persistent storage, a model of a storage medium that is unable to provide any secure deletion of its stored data. After motivating its suitability for modelling remote storage, the chapter then presents a range of related work on the topic of secure deletion for persistently stored data when the user has access to a secondary securely deleting storage medium.

Chapter 9 presents a general approach to the design and analysis of secure deletion for persistent storage that relies on encryption and key wrapping. It defines a key disclosure graph that models the adversarial knowledge over a history of key generation and wrapping. We define a generic update function, expressed as a graph mutation for the key disclosure graph, and prove that this update function achieves secure deletion. Instances of the update function implement the update behaviour of all tree-like data structures including B-Trees, extendible hash tables, linked lists, and others.

Chapter 10 presents a securely deleting data structure using insights from the previous chapter. It uses a B-Tree-based data structure to provide secure deletion. We implement our design in full and analyze its performance, finding that its communication and storage overhead is small.

Chapter 11 considers the problem of an unreliable securely deleting storage medium, that is, one that may lose data, expose data, fail to delete data, and fail to be available. We build a robust fault-tolerant system that uses multiple unreliable storage media. The system permits multiple clients to store securely deletable data and provides a means to control policy aspects of its storage and deletion. It presents details on the implementation both of the distributed securely deleting medium as well as a file system extension that uses it. The solution has low latency at high loads and requires only a small amount of communication among nodes.

Chapter 12 is the conclusive part of this book. We review our contributions and integrate them into our systematization. We present some related and complementary lines of research that fall outside our scope but are still worth discussing. We then outline avenues for future research. Finally, we draw conclusions and summarize our work.

We hope that this book is useful to both academics and practitioners. Those interested in learning about this problem in general and obtaining a thoughtful and novel perspective on it are likely to find Chapters 2, 3, 4, and 8 the most useful: they provide a comprehensive overview of the related work and build a system for comparing and analyzing new solutions based on the properties of solutions made evident in their comparison. As well, researchers may find our own research contributions useful, which are exposed primarily in Chapters 5, 6, 9, and 11. We do not simply devise our systems but further take care to implement and analyze them in practice. As such, the construction and analysis of our systems, presented in Chapters 5, 7, 10 and 11 may be of interest to practitioners. We particularly hope that industry is able to take the flash-based secure deletion software presented in Chapter 7 and implement it in hardware, as a securely deleting memory card would be of great utility and the solution presented here remains the state of the art; Chapter 12 presents a discussion of this.

Chapter 2
Related Work on Secure Deletion

2.1 Introduction

This chapter surveys related work and creates a common language of adversaries, behavioural properties, and environmental assumptions by which to compare and contrast related work. In the next chapter, these concepts are used to design the system and adversarial model that we use throughout this book.

The related work presented in this section provides a background to understand the challenges and complications of secure deletion. It is not comprehensive, however, as further related work specific to secure deletion for flash memory and cloud storage are presented in Section 4.3 and 8.3, respectively.

2.2 Related Work

In this section, we organize related work by the layers through which they access the storage medium. When deciding on a secure deletion solution, one must consider both the interface given to the storage medium and the behaviour of the operations provided by that interface. For example, overwriting a file with zeros uses the file system interface, while destroying the medium uses the physical interface. The solutions available to achieve deletion depend on one's interface to the medium. Secure deletion is typically not implemented by adding a new interface to the storage medium, but instead it is implemented at some existing system layer (e.g., a file system) that offers an interface to the storage medium provided at that layer (e.g., a device driver). It is possible that an interface to a storage medium does not support an implementation of a secure deletion solution.

Once secure deletion is implemented at one layer, then the higher layers' interfaces can explicitly offer this functionality. Care must still be taken to ensure that the secure deletion solution has acceptable performance characteristics: some solutions can be inefficient, cause significant wear, or delete *all* data on the storage medium.

J. Reardon, *Secure Data Deletion*, Information Security and Cryptography,
DOI 10.1007/978-3-319-28778-2_2

These properties are discussed in greater detail in Section 2.4. For now, we first describe the layers and interfaces involved in accessing magnetic hard drives, flash memory, and network file systems on personal computers. We then explain why there is no one layer that is always the ideal candidate for secure deletion. Afterwards, we present related work in secure deletion organized by the layer in which the solution is integrated.

2.2.1 Layers and Interfaces

Many abstraction layers exist between applications that delete data items and the storage medium that stores the data items. Each of these layers may modify how and where data is actually stored.

While there is no standard sequence of layers that encompass all interfaces to all storage media, Figure 2.1 shows the typical ways of accessing flash, magnetic, and networked storage media on a personal computer.

Physical.

The lowest layer is ultimately physical matter—the actual material that can store data and allow access to data, e.g., paper, tape. Its interface is also physical: depending on the medium it can be degaussed, incinerated, or shredded. Additionally, whatever mechanism controls its operation can be replaced with an ad hoc one; for example, flash memory is often accessed through an obfuscating controller, but the raw memory can still be directly accessed by attaching it to a custom reader [26].

Controller.

The storage medium is accessed through a controller. The controller is responsible for translating the data format on the storage media (e.g., electrical voltage) into a format suitable for higher layers (e.g., binary values). Controllers offer a standardized, well-defined hardware interface, such as SCSI or ATA [27], which allows reading and writing to logical fixed-size blocks on the storage medium. They may also offer a *secure erase* command that securely deletes *all* data on the physical device [28]. Like physical destruction, this command cannot be used to securely delete some data while retaining other data; we revisit secure deletion *granularity* later in this chapter.

While hard disk controllers consistently map each logical block to some storage location on the storage medium, the behaviour of other controllers differs. Flash memory does not overwrite existing data; new data is instead logically remapped. When raw flash memory is accessed directly, a different controller interface is exposed. For convenience, flash memory is often accessed through a flash translation

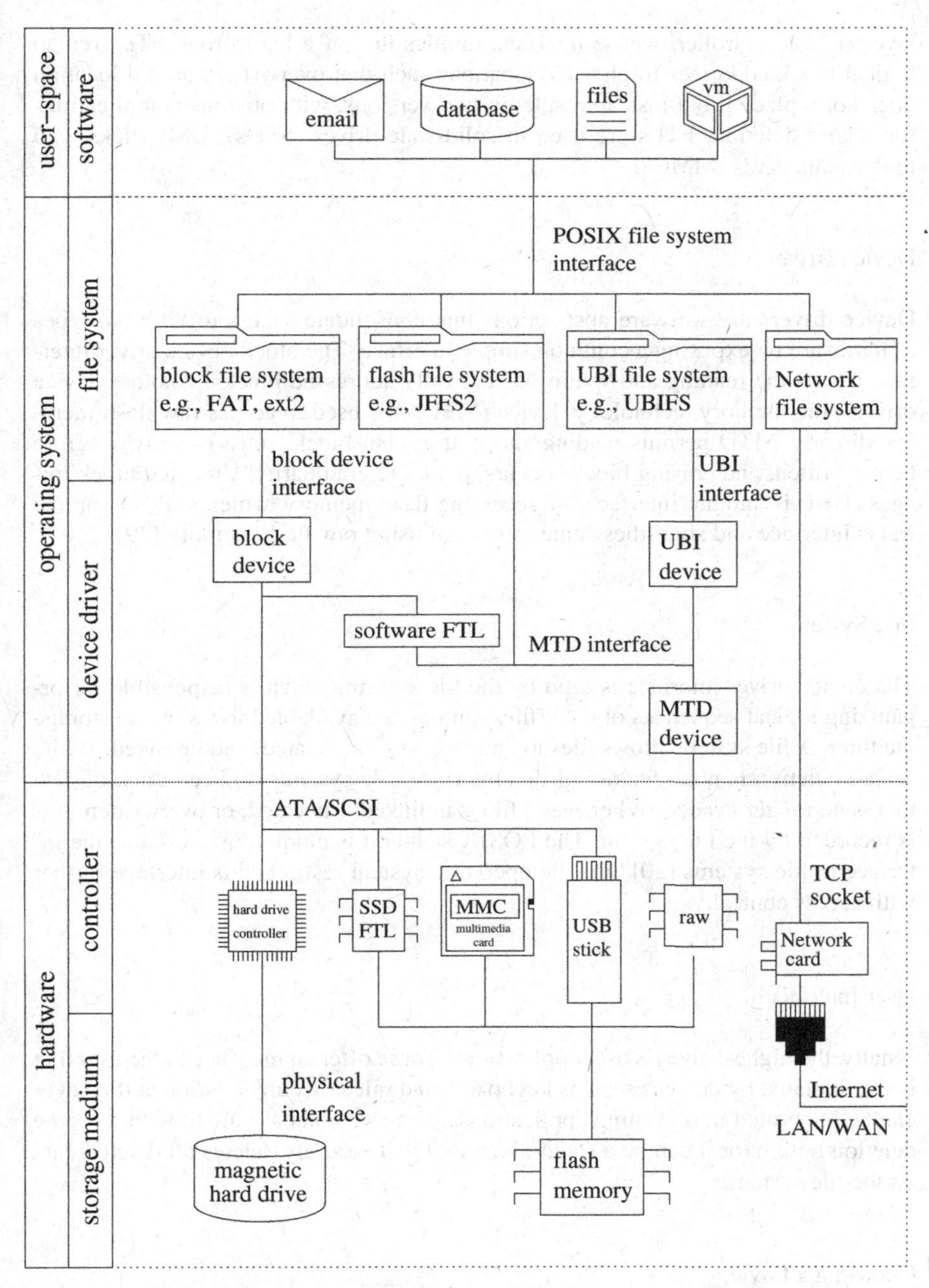

Fig. 2.1 Interfaces and layers involved in magnetic hard drives, flash memory, and remote data storage.

layer (FTL) controller, whose interface mimics that of a hard drive. FTLs remap logical block addresses to physical locations such that overwriting an old location does not replace it but instead results in two versions, with obvious complications for secure deletion. FTLs are used in solid-state drives (SSDs), USB sticks, and multimedia cards (MMCs).

Device Driver.

Device drivers are software abstractions that consolidate access to different types of hardware by exposing a common simple interface. The block device driver interface allows the reading and writing of logically addressed blocks. Another device driver—the memory technology device (MTD)—is used to access raw flash memory directly. MTD permits reading and writing, but blocks must be *erased* before being written, and erasing blocks occurs at a large granularity. Unsorted block images (UBI) is another interface for accessing flash memory, which builds upon the MTD interface and simplifies some aspects of using raw flash memory [29].

File System.

The device driver interface is used by the file system, which is responsible for organizing logical sequences of data (files) among the available blocks on the storage medium. A file system allows files to be read, written, created, and unlinked. While secure deletion is not a feature of this interface, file systems do keep track of data that is no longer needed. Whenever a file is unlinked, truncated, or overwritten, this is recorded by the file system. The POSIX standard is ubiquitously used as the interface to file systems [30], and the operating system restricts this interface further with access control.

User Interface.

Finally, the highest layer is user applications. These offer an interface to the user that is manipulated by devices such as keyboards and mice. Secure deletion at this layer can be integrated into existing applications, such as a database client with a secure deletion option, or it can be a stand-alone tool that securely deletes all deleted data on the file system.

Choosing a Layer.

The choice of layer for a secure-deletion solution is a trade-off between two factors. At the physical layer, we can ensure that the data is truly irrecoverable. At the user layer, we can easily identify the data item to make irrecoverable. Let us consider these factors in more detail.

Each new abstraction layer impedes direct access to the storage medium, thus complicating secure deletion. The controller may write new data, but the storage medium retains remnants; the file system may overwrite a logical block, but the device driver remaps it physically. The further one's interface is abstracted away from the storage medium, the more difficult it is to ensure that one's actions truly result in the irrecoverability of data.

High-layer solutions most easily identify which data items to delete, e.g., by deleting an email or a file. Indirect information is given to the file system, e.g., by unlinking a file; no information is given to the device driver or controller. Assuming the user cannot identify the physical location of the deleted data item on the medium, then a solution integrated at low layers cannot identify where the deleted data item is located. Solutions implemented in the file system are usually well balanced in this trade-off. When this layer is insufficient to achieve secure deletion, it is also possible to pass information on deleted data items from the file system down to lower layers [15, 31].

Organization.

In the remainder of this section, we examine existing secure-deletion solutions organized by the different layers in Figure 2.1. First, we look at *device-level solutions* and *controller-level solutions*, which have no file system information and therefore securely delete all data on the storage medium. We then move to the other extreme and consider *user-level solutions*, which are easy to install and use but are limited in their POSIX-level access to the storage medium and are often rendered ineffective by advanced file system features. We then look at a particular set of *file-system-level solutions*: ones that achieve secure data deletion by overwriting the data that is to be deleted. This technique relies on the device driver actually replacing a location on the storage medium with new content—an *in-place* update—and such updates are not possible for all types of storage media. We conclude with several solutions that extend existing interfaces to allow information on deleted blocks to be sent to lower layers. We defer the survey of techniques suitable for flash memory to Chapter 4, and the survey of encryption-based techniques that are suitable for storing data on persistent storage (e.g., remote "cloud" storage) to Chapter 8.

2.2.2 Physical-Layer and Controller-Layer Sanitization

Physical Layer.

The physical layer's interface is the set of physical actions one can perform on the medium. Secure deletion at this layer often entails physical destruction, but the use of other tools such as degaussers is also feasible. NIST provides physical layer secure-deletion solutions suitable for a variety of types of storage media [11]. For example, destroying floppy disks requires shredding or incineration; destroying compact discs requires incineration or subjection to an optical disk grinding device. Of course, not all solutions work for all media types. For example, most media's physical interfaces permit the media to be put into an NSA/CSS-approved degausser, but this is only a secure deletion solution for particular media types. Magnetic media are securely deleted in this way, while others, such as flash memory, are not.

Controller Layer.

Several standardized interfaces exist for controllers that permit reading and writing of fixed-size blocks. Given these interfaces, there are different actions one can take to securely delete data. Either a single block can be overwritten with a new value to displace the old one, or all blocks can be overwritten. Because the controller has no knowledge of either the deleted data or the organization of data items into blocks, sanitizing a single block cannot guarantee that any particular data item is securely deleted. Therefore, the controller must sanitize every block to achieve secure deletion. Indeed, both SCSI and ATA offer such a sanitization command, called either *secure erase* or *security initialize* [28]. They work like a button that erases all data on the device by exhaustively overwriting every block. The use of these commands is encouraged by NIST as the non-destructive way to securely delete magnetic hard drives. The embedded multimedia card (eMMC) specification allows devices to offer a *sanitize* function. If implemented, it must perform the secure deletion of all data (and copies of data) belonging to *unmapped* storage locations, which includes all data that has been previously marked for deletion.

An important caveat exists at the physical layer. Controllers translate analog values into binary values such that a range of analog values maps to a single binary value. Gutmann observed that, for magnetic media, the precise analog voltage of a stored bit offers insight into its previously held values [32]. Gutmann's solution to deleting this data is also at the controller layer: the controller overwrites every block 35 times with specific patterns designed to ensure analog remnants are removed for all common data encoding formats. In an epilogue to this work, Gutmann remarks that the 35-pass solution had taken on the mystique of a "voodoo incantation to banish evil spirits," and restates that the reason there are so many passes is that it is the concatenation of the passes required for different data encoding formats; it is *never* necessary to perform all 35 for any specific hard drive.

While more recent research was unable to recover overwritten data on modern hard drives [33], it remains safe to say that each additional overwrite does not make the data easier to recover—in the worst case it simply provides no additional benefit [6]. Gutmann's epilogue states that it is unlikely anything can be recovered from modern drives after a couple of passes with random data. More generally, Gutmann's results highlight that analog remnants introduced by the controller's use of the storage medium may exist for any storage media type and this must be considered when developing secure-deletion solutions.

2.2.3 User-Level Solutions

Device-level solutions interact at the lowest layer and securely delete all data, serving as a useful starting point in our systematization. Now we move to the other extreme, a securely deleting user-level application that can only interact with a POSIX-compliant file system. There are three common user-level solutions: (i) ones that call a secure deletion routine in the storage medium's interface, (ii) ones that overwrite data before unlinking, and (iii) ones that first unlink and then fill the empty capacity of the storage medium.

Low-Layer Calls.

Device drivers and other low-layer interfaces may expose to user-space special routines for secure deletion. This permits users to easily invoke such functionality without requiring special access to hardware or additional skills. Explicit low-layer calls propagate a secure-deletion solution to a higher-layer interface.

Hughes et al. offer a free Secure Erase utility [28]. It is a user-level application that securely erases all data on a storage medium by invoking the Secure Erase command in the hardware controller's interface.

Similarly, Linux's MMC driver exposes to user-space an *ioctl* that invokes the sanitization routine [34]. Therefore, applications can easily call the ioctl, which—if supported by the hardware—performs the secure deletion of all unmapped data on the storage medium.

File Overwriting Tools.

Another class of user-level secure-deletion solutions opens up a file from user-space and *overwrites* its contents with new, insensitive data, e.g., all zeros. When the file is later unlinked, only the contents of the most-recent version are currently stored on the storage medium. To combat analog remnants, overwriting is performed multiple times; multiple tools [35, 36] offer the 35-pass overwriting as proposed by Gutmann [32].

Overwriting tools rely on the following file system property: each file block is stored at known locations and when the file block is updated, then all old versions are replaced with the new version. If this assumption is not satisfied, user-level overwriting tools silently fail. Moreover, they do nothing for larger files that were truncated at some time prior to running the tool. They also do nothing for file copies that are not unlinked with this tool.

Overwriting tools may also attempt to overwrite file system metadata, such as its name, size, and access times. The Linux tool `wipe` [36], for instance, also changes the file name and size in an attempt to securely delete this metadata. Note that not all types of metadata may be arbitrarily changed: the operating system's interface to the file system may not allow it, or simply changing the file's name, for example, may not securely delete the old one. The Linux tool `srm` [35] renames the file to a random value and truncates its size to zero after overwriting. Other attributes cannot be easily changed without higher privileges, e.g., the access times or the file's group and owner.

Overwriting tools operate on either an entire file or the entire storage medium. These tools do not handle operations such as overwrites and truncations, which discard data within a file without deleting the file. Though overwriting a file replaces the old data with the new data (or else such an overwriting tool is unsuitable), it does not perform additional sanitization steps such as writing over the location multiple times. While it is possible to write a user-level tool that securely overwrites and truncates a file as well, it becomes the user's burden to ensure that all other applications make use of it.

This leads into the general problem of usability. The user must remember to use the tool whenever a sensitive file must be deleted, and to do this instead of their routine behaviour. Care must be taken to avoid applications that create and delete their own files [16]: a word processor that creates temporary swap files does not securely delete them with the appropriate tool; a near-exact copy is left available. If a file is copied, the copy too must be securely deleted. Neglecting to use the tool when deleting a file results in the inability to securely delete the file's data with this technique.

Free-Space Filling Tools.

A file system has both valid and unused blocks. The set of unused blocks is a superset of the blocks containing deleted sensitive data. A third class of user-level secure-deletion tools exploits this fact by *filling* the entire free space of the file system. This ensures that *all* unused blocks of the storage medium no longer contain sensitive information and instead store filler material.

Filling solutions permit users and applications to take no special actions to securely delete data; any discarded data is later securely deleted by an intermittent filling operation. These tools also allow secure deletion for file systems that do not perform in-place updates of file data. Compared to the overwriting solutions, secure deletion through filling allows *per-block-level* secure deletion (including truncations) without in-place updates at the cost of a periodic operation. It can only operate at *full scope*—all unused blocks are filled. Examples include Apple's Disk Utility's *erase free space* feature [37] and the open-source tool `scrub` [38].

The correct operation of a filling tool relies on two assumptions: the user who runs the tool must have sufficient privileges to fill the storage medium to capacity, and when the file system reports itself as unwritable it must no longer contain any deleted data. The useful deployment of these solutions therefore requires manual verification that these assumptions do hold.

Filling's correctness assumptions are satisfied more often than overwriting's correctness assumptions. Modern file systems, for example, typically do not overwrite data in place but instead use journalling when storing new data for crash-recovery purposes. Filling's assumptions, however, also do not always hold. Garfinkel and Malan examine secure deletion by filling for a variety of file systems and find mixed results [39]. One observation is that creating many small files helps securely delete data that is not deleted when creating one big file, which may be due to file systems not allocating heavily fragmented areas for already large files.

The benefits of filling over overwriting are that the user is given secure deletion for all deleted data (including unmarked sensitive files and truncations) that works correctly for a larger set of file systems. Moreover, the user only needs to run the tool periodically to securely delete all accumulated deleted data: applications and user behaviour do not need to change with regards to file management. The trade-off is that the filling operation is slow and cannot target specific files. It is a periodic operation that securely deletes all data blocks discarded within the last period. Since deletion is achieved by writing new data instead of overwriting the existing data, it does not perform in-place updates and is therefore suitable for additional file systems and storage medium types that do not permit such operations.

Database Secure Deletion.

Databases such as MySQL [40] and SQLite [41] store an entire database as a single file within a file system [42]; databases are analogous to file systems, where records can be added, removed, and updated. This adds a new interface layer for users wanting to delete entries from a database. Database files are long lived on a system; the data they contain, however, may only be valid for a short time. Many applications store sensitive user data (e.g., emails and text messages) in databases; secure deletion of such data from databases is therefore important.

Both MySQL and SQLite have secure-deletion features. In both cases, the interface for secure deletion is the underlying file system and secure deletion is implemented with in-place updates. For MySQL, researchers propose a solution where deleted entries are overwritten with zeros, and the transaction log (used to recover after a crash) is encrypted and securely deleted by deleting the encryption key [42]. For SQLite, there is an optional secure-deletion feature that overwrites deleted database records with zeros [43].

As previously discussed, overwriting blocks with zeros is one way to inform the file system that these blocks are unneeded—necessary, but not sufficient, to achieve secure deletion. SQLite's solution relies on the file system "below" to ensure that overwritten data results in its secure deletion. When the interface does not explicitly offer secure deletion, then it is—at the minimum—necessary to tell the interface that the data is discarded.

2.2.4 File-System-Level Solutions with In-Place Updates

The utility of user-level solutions is hampered by the lack of direct access to the storage medium. Device-level solutions suffer from being generally unable to distinguish deleted data from valid data given that they lack the semantics of the file system. We now look at secure-deletion solutions integrated in the file system itself, that is, solutions that access the storage medium using the device driver interface.

Here we consider only solutions that use in-place updates to achieve secure deletion. An in-place update means that the device driver replaces a location on the storage medium with new content. Not all device drivers offer this in their interface, primarily because not all storage media support in-place updates. The assumption that in-place updates occur is valid for block device drivers that access magnetic hard drives and floppy disks. Solutions for flash memory cannot use in-place updates, and Section 4.3 discusses this in more detail.

Secure Deletion for ext2.

The second extended file system ext2 [44] for Linux offers a sensitive attribute for files and directories to indicate that secure deletion should be used when deleting the file. While the actual feature was never implemented by the core developers, researchers provided a patch that implements it [16].

Their patch changed the functionality that marks a block as free. It passes freed blocks to a kernel daemon that maintains a list of blocks that must be sanitized. If the free block corresponds to a sensitive file, then the block is added to the work queue instead of being returned to the file system as an empty block. The work queue is sorted to minimize seek times imposed by random access on spinning-disk magnetic media.

The sanitization daemon runs asynchronously, performing sanitization when the system is idle, allowing the user to perceive immediate file deletion. The actual sanitization method used is configurable, from a simple overwrite to repeated overwrites in accordance with various standards.

Secure Deletion for ext3.

The third extended file system ext3 [45] succeeded ext2 as the main Linux file system and extended it with a write journal: all data is first written into a journal before being committed to main storage. This improves consistent state recovery after unexpected losses of power by only needing to inspect the journal's recent changes.

Joukov et al. [46] provide two secure-deletion solutions for ext3. Their first solution is a small change that provides secure deletion of file data by overwriting it once, which they call *ext3 basic*. Their second solution, *ext3 comprehensive*, provides secure deletion of file data and file metadata by overwriting it using a configurable overwriting scheme, such as the 35-pass Gutmann solution. They both provide secure deletion for all data or just those files whose extended attributes include a sensitive flag.

Secure Deletion via Renaming.

Joukov et al. [46] present another secure-deletion solution through a file system extension, which can be integrated into many existing file systems [47]. Their extension intercepts file system events relevant for secure deletion: unlinking a file and truncating a file. (They assume overwrites occur in place and are not influenced by a journal or log-structured file system.) For unlinking, which corresponds to regular file deletion, their solution instead moves the file into a special secure-deletion directory. For truncation, the resulting truncated file is first copied to a new location and the older, larger file is then moved to the special secure-deletion directory. Thus, for truncations, their solution must always process the entire file—not just the truncated component. At regular intervals, a background process runs the user-level tool `shred` [48] on all the files in the secure-deletion directory.

Purgefs.

Purgefs is another file system extension that adds secure deletion to any block-based file system [49]. It uses block-based overwriting when blocks are returned to the file system's free list, similar to the solution used for ext2 [44]. It supports overwriting file data and file metadata for all files or just files marked as sensitive. Purgefs is implemented as a generic file system extension, which can be combined with any block-based file system to create a new file system that offers secure deletion.

Secure Deletion for a Versioning File System.

A versioning file system shares file data blocks among many versions of a file; one cannot overwrite the data of a particular block without destroying all versions that share that block. Moreover, user-level solutions such as overwriting the file fail to securely delete data because all file modifications are implemented using a copy-on-write semantics [50]—a copy of the file is made (sharing as many blocks as possible with older versions) with a new version for the block now containing only zeros.

Peterson et al. [51] use a cryptographic solution to optimize secure deletion for versioning file systems. They use an all-or-nothing cryptographic transformation [52] to expand each data block into an encrypted data block along with a small key-sized tag that is required to decrypt the data. If any part of the ciphertext is deleted—either the tag or the message—then the entire message is undecipherable. Each time a block is shared with a new version, a new tag is created and stored for that version. Tags are stored sequentially for each file in a separate area of the file system to simplify sequential access to the file under the assumption that a magnetic-disk drive imposes high seek penalties for random access. A specific version of a file can be quickly deleted by overwriting all of that version's tags. Moreover, *all* versions of a particular data block can easily be securely deleted by overwriting the encrypted data block itself.

2.2.5 Cross-layer Solutions

There are solutions that pass information on discarded data down through the layers, permitting the use of efficient low-layer secure-deletion solutions.

Data items contained in a file are discarded from a file system in three ways: by unlinking the file, by truncating the file past the block, and by updating the data item's value. The information about data blocks that are discarded when unlinking or truncating files, however, remains known only by the file system. The device-driver layer can only infer the obsolescence of an old block when its logical address is overwritten with a new value. Here we present two solutions by which the file system passes information on discarded blocks to the device driver: TRIM commands [31] and TrueErase [15]. In both cases, the file system informs the device that particular blocks are discarded, i.e., no longer needed for the file system. With this information, the device driver can implement its own efficient secure deletion without requiring data blocks to be explicitly overwritten by the file system.

TRIM commands are notifications issued from the file system to the device driver to inform the latter about discarded data blocks. TRIM commands were not designed for secure deletion but instead as an efficiency optimization for flash-based storage media. Nevertheless, there is no reason that a device driver cannot use information from TRIM commands to perform secure deletion: TRIM commands indicate every time a block is discarded—there are no false negatives. It is not possible to restrict TRIM commands only to sensitive blocks, which means that it must be an efficient underlying mechanism that securely deletes the data.

Diesburg et al. propose TrueErase [15], which provides similar information as TRIM commands but only for blocks belonging to files specifically marked as sensitive. Users may simply set all files to sensitive or use traditional permission semantics to manage file sensitivity. TrueErase adds a new communication channel between the file system and the device driver that forwards from the former to the latter information on sensitive blocks deleted from the file system. Device drivers are modified to implement immediate secure deletion when provided a deleted block; the device driver is thus able to correctly implement secure deletion using its lower-layer interface with the high-layer information on what needs to be deleted. This is more efficient than TRIM commands, which would require deletion for all data. This comes at the risk of a false negative in the event that a user neglects to correctly set a file's sensitivity.

2.2.6 Summary

This concludes our survey of selected related work on secure deletion. Further related work specific to flash memory and persistent memory appear in Chapters 4 and 8, respectively. We saw that storage media can be accessed from a variety of layers and that different layers provide different interfaces for secure deletion. In low-layer solutions, fewer assumptions must be made about the interface's behaviour, while in high-layer solutions the user can most clearly mark which data items to delete. For device-level solutions, we discussed different ways the entire device can be sanitized. User-level secure deletion considers how to securely delete data using a POSIX-compliant file system interface. Secure deletion in the file system must use the device driver's interface for the storage medium, and we surveyed solutions that assume the device driver performs in-place updates. For storage media that do not have an erasure operation, physical destruction is the only means to achieve secure deletion.

In the next two sections, we organize the space of secure-deletion solutions. We first review adversarial models and afterwards compare the characteristics of existing solutions.

2.3 Adversarial Model

Secure-deletion solutions must be evaluated with respect to an adversary. The adversary's goal is to recover deleted data items after being given some access to a storage medium that contained some representation of the data items. In this section, we present the secure-deletion adversaries. We develop our adversarial model by abstracting from real-world situations in which secure deletion is relevant, and identifying the classes of adversarial capabilities characterizing these situations. Table 2.1 then presents a variety of real-world adversaries organized by their capabilities.

2.3.1 Classes of Adversarial Capabilities

Attack Surface.

The attack surface is the storage medium's interface given to the adversary. If deletion is performed securely, data items should be irrecoverable to an adversary who has unlimited use of the provided interface. NIST divides the attack surface into two categories: *robust-keyboard attacks* and *laboratory attacks* [11]. Robust-keyboard attacks are software attacks: the adversary acts as a device driver and accesses the storage medium through the controller. Laboratory attacks are hardware attacks: the adversary accesses the storage medium through its physical interface. As we have seen, the physical layer may have analog remnants of past data inaccessible at any other layer. While these two surfaces are widely considered in related work, we emphasize that any interface to the storage medium can be a valid attack surface for the adversary.

Access Time.

The access time is the time when the adversary obtains access to the medium. Many secure-deletion solutions require performing extraordinary sanitization methods before the adversary is given access to the storage medium. If the access time is unpredictable, the user must rely on secure deletion provided by sanitization methods executed as a matter of routine.

The access time is divided into two categories: predictable (or user controlled) and unpredictable (or adversary controlled). If the access time is predictable, then the user can use the storage medium normally and perform as many sanitization procedures as desired before providing it to the adversary. If the access time is unpredictable then we do not permit any extraordinary sanitization methods to be executed: the secure-deletion solution must rely on some immediate or intermittent sanitization operation that limits the duration that deleted data remains available.

Number of Accesses.

Nearly all secure-deletion solutions consider an adversary who accesses a storage medium some time after securely deleting the data. One may also consider an adversary who accesses the storage medium multiple times—accessing the storage medium before the data is written as well as after it is deleted, similar to the *evil maid* attack on encrypted file systems.

We therefore differentiate between single- and multiple-access adversaries. A single-access adversary may surface when a used storage medium is sold on the market; a multiple-access adversary is someone who, for example, deploys malware on a target machine multiple times because it is discovered and cleaned, or someone who obtains surreptitious periodic access (e.g., nightly access) to a storage medium.

Credential Revelation.

Encrypting data makes it immediately irrecoverable to an adversary that neither has the encryption key (or user passphrase) nor can decrypt data without the corresponding key. There are many situations, however, where the adversary is given this information: a legal subpoena, border crossing, or information taken from the user through duress. In these cases, encrypting data is insufficient to achieve secure deletion.

We partition the credential revelation into non-coercive and coercive adversaries. A non-coercive adversary does not obtain the user's passwords and the credentials that protect the data on the storage medium. A coercive adversary, in contrast, obtains this information. It may also be useful to consider a weak-password adversary who can obtain the user's password by guessing, by the device not being in a locked state, or by a cold-boot attack [53]. This adversary, however, is unable to obtain secrets such as the user's long-term signing key or the value stored on a two-factor authentication token.

Computational Bound.

Many secure-deletion solutions rely on encrypting data items and only storing their encrypted form on the medium. The data is made irrecoverable by securely deleting the decryption key. The security of such solutions must assume that the adversary is computationally bounded to prevent breaking the cryptography.

We distinguish between computationally bounded and unbounded adversaries. There is a wealth of adversarial bounds corresponding to a spectrum of non-equivalent computational hardness problems, so others may benefit from dividing this spectrum further.

2.3.2 Summary

Adversaries are defined by their capabilities. Table 2.1 presents a subset of the combinatorial space of adversaries that correspond to real-world adversaries. The *name* column gives a name for the adversary, taken from related work when possible; this adversarial name is later used in Table 2.3 (and much later in Table 12.1) when describing the adversary a solution defeats. The second through fifth columns correspond to the classes of capabilities defined in this section. Table 2.2 provides a real-world example where each of the adversaries from Table 2.1 may be found. For instance, while computationally unbounded adversaries do not really exist, the consideration of such an adversary may reflect a corporate policy on the export of sensitive data or a risk analysis of a potentially broken cryptographic scheme.

Observe that each class of adversarial capabilities is *ordered* based on adversarial strength: lower-layer adversaries get richer data from the storage medium, coercive adversaries get passwords in addition to the storage medium, and an adversary who controls the disclosure time can prevent the user from performing an additional extraordinary effort to achieve secure deletion. This yields a partial order on adversaries, where an adversary A is *weaker than or equal to* an adversary B if all of A's capabilities are weaker than or equal to B's capabilities. A is strictly *weaker* than B if all of A's capabilities are weaker than or equal to B's and at least one of A's capabilities is weaker than B's. Consequently, a secure-deletion solution that defeats

Table 2.1 Taxonomy of secure-deletion adversaries.

Adversary's Name	Disclosure	Credentials	Bound	Accesses	Surface
internal repurposing	predictable	non-coercive	bounded	sing/mult	controller
external repurposing	predictable	non-coercive	bounded	single	physical
advanced forensic	predictable	non-coercive	unbounded	single	physical
border crossing	predictable	coercive	bounded	sing/mult	physical
unbounded border	predictable	coercive	unbounded	sing/mult	physical
malware	unpredict.	non-coercive	bounded	sing/mult	user-level
compromised OS	unpredict.	either	bounded	sing/mult	block dev
bounded coercive	unpredict.	coercive	bounded	single	physical
unbounded coercive	unpredict.	coercive	unbounded	single	physical
bounded multi-access	unpredict.	coercive	bounded	multiple	physical
unbounded multi-access	unpredict.	coercive	unbounded	multiple	physical

Table 2.2 Example of adversaries modelled.

Adversary's Name	Example
internal repurposing	loaning hardware
external repurposing	selling old hardware
advanced forensic	unfathomable forensic power
border crossing	perjury to not reveal password
unbounded border crossing	cautious corporate policy on encrypted data
malware	malicious application
compromised OS	operating system malware, passwords perhaps provided
bounded coercive	legal subpoena
unbounded coercive	legal subpoena and broken crypto
bounded multiple-access	legal subpoena with earlier spying
unbounded multiple-access	legal subpoena, spying, broken crypto

an adversary also defeats all weaker adversaries, under this partial ordering. Finally, as expected, A is *stronger* than B if B is weaker than A.

2.4 Analysis of Solutions

Secure-deletion solutions have differing characteristics, which we divide into assumptions on the environment and behavioural properties of the solution. *Environmental assumptions* include the expected behaviour of the system underlying the interface; *behavioural properties* include the deletion latency and the wear on the storage medium. If the environmental assumptions are satisfied then the solution's behavioural properties should hold, the most important of which is that secure deletion occurs. No guarantee is provided if the assumptions are violated. It may also be the case that stronger assumptions yield solutions with improved properties.

In this section, we describe standard classes of assumptions and properties. Table 2.3 organizes the solutions from Section 2.2 into this systematization.

2.4.1 *Classes of Environmental Assumptions*

Adversarial Resistance.

An important assumption is the one made on the strength and capabilities of the adversary, as defined in Section 2.3. For instance, a solution may only provide secure deletion for computationally bounded adversaries; the computational bound is an assumption required for the solution to work. A solution's adversarial resistance is a set of adversaries; adversarial resistance assumes that the solution need not defeat any adversary stronger than an adversary in this set.

System Integration.

This chapter organizes the secure-deletion solutions by the interface through which they access the storage medium. The interface that a solution requires is an environmental assumption, which assumes that this interface exists and is available for use. System integration may also include assumptions on the behaviour of the interface with regards to lower layers (e.g., that overwriting a file securely deletes the file at the block layer). For instance, a user-level solution assumes that the user is capable of installing and running the application, while a file-system-level solution assumes that the user can change the operating system that accesses the storage medium. The ability to integrate solutions at lower-layer interfaces is a stronger assumption than at higher layers because higher-layer interfaces can be simulated at a lower layer.

System integration also makes assumptions about the interface's behaviour. For example, solutions that overwrite data with zeros assume that this operation actually replaces all versions of the old data with the new version. When such interface assumptions are not satisfied, then the solution does not provide secure deletion. Table 2.3 notes the solutions that require in-place updates in order to correctly function.

2.4.2 *Classes of Behavioural Properties*

Deletion Granularity.

The granularity of a solution is the solution's deletion unit. We divide granularity into three categories: *per storage medium*, *per file*, and *per data item*. A per-storage-medium solution deletes *all* data on a storage medium. Consequently, it is an extraordinary measure that is only useful against a *user-controlled access time* adversary, as otherwise the user is required to completely destroy all data as a matter of routine. At the other extreme is sanitizing deleted data at the smallest granularity offered by the storage medium: e.g., block size, sector size, or page size. Per-data-item solutions securely delete any deleted data from the file system, no matter how small.

Between these extremes lies per-file secure deletion, which targets files as the deletion unit: a file remains available until it is securely deleted. While it is common to reason about secure deletion in the context of files, we caution that the file is not the natural unit of deletion; it often provides similar utility as per-storage-medium deletion. Long-lived files such as databases frequently store user data; the Android phone uses them to store text messages, emails, etc. A virtual machine may store an entire file system within a file: anything deleted from this virtual file system remains until the user deletes the entire virtual machine's storage medium. In such settings, per-file secure deletion requires the deletion of all stored data in the DB or VM, which is an extraordinary measure unsuitable against adversaries who control the disclosure time. In other settings, such as the storage of large media files, file data tends to be stored and deleted at the granularity of an entire file and so per-file solutions may reduce overhead.

Scope.

Many secure-deletion solutions use the notion of a sensitive file. Instead of securely deleting all deleted data from the file system in an untargeted way, they only securely delete known sensitive files, and require the user to mark sensitive files as such. We divide the solution's scope into *untargeted* and *targeted*. A targeted solution only securely deletes sensitive files and can substitute for an untargeted solution simply by marking every file as sensitive.

While targeted solutions are more efficient than untargeted ones, we have some reservations about their usefulness. First, the file is not necessarily the correct unit to classify data's sensitivity; an email database is an example of a large file whose content has varying sensitivity. The benefits of targeting therefore depend on the deployment environment. Second, some solutions do not permit files to be marked as sensitive after their initial creation, such as solutions that must encrypt data items *before* writing them onto a storage medium. Such solutions are not suitable for cases where, for example, users manually mark emails from the inbox as sensitive so that additional secure-deletion actions are taken when it is later deleted. Finally, targeted solutions introduce usability concerns and consequently false classifications due to user error. Users must take deliberate action to mark files as sensitive. A false positive costs efficiency while a false negative may disclose confidential data. While usability can be improved with a tainting-like strategy for sensitivity [54], this is still prone to erroneous labelling and requires user action. Previous work has shown the difficulty of using security software correctly [55] (even the concept of a deleted items folder retaining data confounds some users [56]) and security features that are too hard to use are often circumvented altogether [6].

A useful middle ground is to broadly partition the storage medium into a securely deleting user-data partition and a normal operating system partition. Untargeted secure deletion is used on the user-data partition to ensure that there are no false negatives and this requires no change in user behaviour or applications. No secure deletion is used for the OS partition to gain efficiency for files deemed not sensitive.

Device Lifetime.

Some secure-deletion solutions incur device wear. We divide device lifetime into complete wear, some wear, and unchanged. *Complete wear* means that the solution physically destroys the medium. *Some wear* means that a non-trivial reduction in the medium's expected lifetime occurs, which may be further subdivided with finer granularity based on notions of wear specific to the storage medium. *Unchanged* means that the secure-deletion operation has no significant effect on the storage medium's expected lifetime.

Deletion Latency.

Secure-deletion latency refers to the timeliness when secure-deletion guarantees are provided. There are many ways to measure this, such as how long one expects to wait before deleted data is securely deleted. Here, we divide latency into immediate and periodic secure deletion.

An *immediate* solution is one whose deletion latency is negligibly small. The user is thus assured that data items are irrecoverable promptly after their deletion. This includes applications that immediately delete data as well as file system solutions that only need to wait until a kernel sanitization thread is scheduled for execution.

A *periodic* solution is one that intermittently executes and provides a larger deletion latency. Such solutions, if run periodically, provide a fixed worst-case upper bound on the deletion latency of all deleted data items. Periodic solutions involve batching: collecting many pieces of deleted data and securely deleting them simultaneously. This is typically for efficiency reasons. An important factor for periodic solutions is crash-recovery. If data items are batched for deletion between executions and power is lost, then either the solution must recover all the data to securely delete when restarted (e.g., using a commit and replay mechanism) or it must securely delete all deleted data without requiring persistent state (e.g., filling the hard drive [36, 38, 57]).

Table 2.3 Spectrum of secure-deletion solutions

Solution Name	Target Adversary	Integration	Granularity	Scope
overwrite [35, 36, 48]	unbounded coercive	user-level[a]	per-file	targeted
fill [37, 38, 57]	unbounded coercive	user-level	per-data-item	untargeted
NIST clear [11]	internal repurposing	varies	per-medium	untargeted
NIST purge [11]	external repurposing	varies	per-medium	untargeted
NIST destroy [11]	advanced forensic	physical	per-medium	untargeted
ATA secure erase [28]	external repurposing	controller	per-medium	untargeted
renaming [46]	unbounded coercive	kernel[a]	per-data-item	targeted
ext2 sec del [16]	unbounded coercive	kernel[a]	per-data-item	targeted
ext3 basic [46]	unbounded coercive	kernel[a]	per-data-item	targeted
ext3 comprehensive [46]	unbounded coercive	kernel[a]	per-data-item	targeted
purgefs [49]	unbounded coercive	kernel[a]	per-data-item	targeted
ext3cow sec del [51]	bounded coercive	kernel[a]	per-data-item	untargeted

[a] Assumes interface performs in-place updates.

Solution Name	Lifetime	Latency	Efficiency
overwrite [35, 36, 48]	unchanged	immediate	number of overwrites
fill [37, 38, 57]	unchanged	immediate	depends on medium size
NIST clear [11]	varies	immediate	varies with medium type
NIST purge [11]	varies	immediate	less efficient than clearing
NIST destroy [11]	destroyed	immediate	varies with medium type
ATA secure erase [28]	unchanged	immediate	depends on medium size
renaming [46]	unchanged	immediate	truncations copy the file
ext2 sec del [16]	unchanged	immediate	batches to minimize seek
ext3 basic [46]	unchanged	immediate	batches to minimize seek
ext3 comprehensive [46]	unchanged	immediate	slower then ext3 basic
purgefs [49]	unchanged	immediate	number of overwrites
ext3cow sec del [51]	unchanged	immediate	deletes multiple versions

Efficiency.

Solutions often differ in their efficiency. Wear and deletion latency are two efficiency metrics we explicitly consider. The particular relevant metrics depend on the application scenario and the storage medium. Other metrics include the ratio of bytes written to bytes deleted, battery consumption, storage overhead, execution time, etc. The metric chosen depends on the underlying storage medium and use case.

2.4.3 Summary

Table 2.3 presents the spectrum of secure-deletion solutions organized into the framework developed in this section. For brevity, we do not list all adversaries that a solution can defeat, but instead state what we inferred as the solution's target adversary.

The classes of environmental assumptions and behavioural properties are each ordered based on increased *deployment requirements*. Solutions that cause wear and use in-place updates have stronger deployment requirements (i.e., that wear is permitted, and the interface allows and correctly implements in-place updates) than solutions that do not cause wear or use in-place updates. Solutions that defeat weak adversaries have stronger deployment requirements (i.e., that the adversary is weak) than solutions that defeat stronger adversaries. The result is a partial ordering of solutions that reflects substitutability: a solution with weaker deployment requirements can replace one with stronger deployment requirements as it requires less to correctly deploy.

Chapter 3
System Model and Security Goal

3.1 Introduction

This chapter presents the system model, adversarial model, and security goal for the work presented in this book. First, the system model describes how the user interacts with storage media to store, read, and delete data. Second, the storage medium models describe a number of different storage media types relevant to this book. Third, the adversarial model describes how the adversary is able to gain access to the user's storage media. Finally, the security goal defines secure deletion and describes what data our solutions strive to delete.

3.2 System Model

Our system model consists of a user who stores data on storage media such that the data can be retrieved by the user during the data's lifetime. A data's *lifetime* is the range between two events: the data's initial *creation* and its subsequent *discard*. Data is considered *valid* during its lifetime and *invalid* or *not valid* at all other times.

We assume that the user divides the data to store into discrete *data items* that share a lifetime. These can be binary data objects, files, or individual blocks of a block-based file system. The user retains a *handle* to recall these items (e.g., an object name or a block address). The set of handles and the mapping of handles to storage locations may entail the storage of metadata. We do not elevate metadata to require particular concern: metadata is itself stored as data items and can therefore be securely deleted in the same way.

The user continuously stores, reads, and discards data on a storage medium. For example, it may be a mobile phone storing location data continually throughout the day, or a server continually storing sensitive log data needed only for a short time to monitor for malicious behaviour. The user may use multiple storage media when storing and retrieving data. These media may differ in their implementation

J. Reardon, *Secure Data Deletion*, Information Security and Cryptography,
DOI 10.1007/978-3-319-28778-2_3

Table 3.1 Mapping the simple storage interface to different storage interfaces. Linux function names or constant definitions for standardized hardware signals are used for the interface.

Storage Interface (IX)	Handles	Contents
Object Store (OSD) [58]	object id	object
Distributed Hash Table (DHT) [60]	key	value
POSIX filesystem (FS) [30]	file and offset	file contents
Block device (blk)	block address	block
ATA device (ATA) [27]	logical block address	sector
Multimedia card (MMC) [61]	data address	block

		Simple storage interface	
IX	**Store**	**Read**	**Discard**
OSD	create, write	read	remove
DHT	store	find value	-
FS	write	read	truncate, unlink
blk	REQ_WRITE	REQ_READ	REQ_DISCARD
ATA	ATA_CMD_WRITE	ATA_CMD_READ	ATA_DSM_TRIM
MMC	MMC_DATA_WRITE	MMC_DATA_READ	MMC_TRIM_ARG

and interface. We assume that the user can store data using an object store device interface (e.g., OSD [58, 59]): the user can *store* data (an object) with a handle (lookup key), *read* the data for a handle, and *discard* the data for a handle. The use of the nomenclature *discard* is intentional to emphasize that discard does not necessarily entail any deletion.

Particular storage media may have expanded interfaces, but we assume that the user stores, reads, and discards data with the object store interface. Updating data is implemented by discarding the old version and storing a new version. Securely deleting data is achieved when discarded data is irrecoverable from a storage medium. Table 3.1 describes how these three storage functions translate into a variety of common storage interfaces.

A fundamental correctness property for a storage medium is that it retrieves stored data. Particularly, from the moment data is stored until it is later discarded, the data must be readable. This property makes no requirements on data not being readable after it is discarded (or for that matter, before it is created); it only requires that data is readable when it is considered valid.

3.3 Storage Medium Models

We now describe a suite of storage medium models representative of a variety of real-world systems and which differ greatly in how easily data can be securely deleted. At the extremes, we have the secure-deletion optimal SECDEL model and the secure-deletion near-pessimal PERSISTENT model.

Securely Deleting Model.

The SECDEL model is an idealized case where the interface's deletion function performs immediate secure deletion. This models any black-box-like storage system that correctly implements secure deletion. Secure deletion solutions transform other media or ensembles of media into a SECDEL-like model through an explicit construction.

An analog example is a rolling index of numbered cards: data is written onto new cards, which are then numbered by the position and inserted into the index; discarding removes the numbered card and incinerates it.

Clocked Securely Deleting Models.

The two clocked models—SECDEL-CLOCK and SECDEL-CLOCK-EXIST—are idealized cases that model a storage system that correctly implements a periodic secure-deletion operation. We include these idealization because some secure-deletion solutions have non-trivial execution costs and are therefore run periodically to compensate. Clocked models have a *clock period*, that is, the time between each *clock edge*. The clock divides time into discrete *deletion epochs*, where all data discarded in one deletion epoch is securely deleted at the clock edge and therefore no longer readable in later deletion epochs.

In SECDEL-CLOCK, discarded data remains stored until the next clock edge; in SECDEL-CLOCK-EXIST, discarded data remains stored on the medium until the next clock edge in *both directions*: before its creation and after its discard.

An analog example of SECDEL-CLOCK is paper recycling in a security-conscious organization. Data written on paper is discarded into recycling bins. Each night, all scrap paper from recycling bins is shredded before recycling.

In-Place Update Model.

The INPLACE model is does not securely delete data when it is discarded; instead, the corresponding position is marked for deletion, which indicates that its consumed resources can be reclaimed when needed. This models simple file systems such as FAT and ext2 as well as the block device interface that accesses magnetic storage media such as hard drives and floppy disks. Much of the related work presented in Section 2.2 makes the assumption that data is stored on an in-place update storage medium.

An analog example is a set of blackboards during a lecture. New data is written on a blank board, but old data is not securely deleted until the space it consumes is needed for new data.

Semi-persistent Model.

The SEMIPERSISTENT model is a finite-size medium where the interface's concept of a storage position differs from the implementation's concept. An indirection layer maps interface positions to storage positions. The user may store data at logical positions but is given no control over where data is physically stored. The semi-persistent nature of the medium is a consequence of its finite size, necessitating the eventual reuse of physical positions and therefore the secure deletion of previous content. This model corresponds to a variety of log-structured file systems such as JFFS, YAFFS, and UBIFS as well as the effect of accessing flash memory through a flash translation layer (FTL).

An analog example is an inter-office mail envelope; old recipients' names remain visible but the current valid recipient is the one furthest down. When a name is written in all the slots, the names are erased by discarding the envelop and replacing it.

Persistent Model.

The PERSISTENT model is a storage medium that does not ever securely delete data. Once data is stored on the persistent storage it remains stored permanently. This may occur because the nature of the medium is indelible and append-only, such as writing information on optical discs and storing them in an archive. The medium may also not explicitly be append only, but environmental assumptions warrant it to be considered as such; for example, an adversary that obtains continuous access to stored data (i.e., by monitoring the network or controlling the storage medium), or a user who is unable to gauge the adversary's eventual forensic capabilities.

An analog example is publishing data in a newspaper; corrections can be issued for incorrect data, however the published data remains a persistent part of the public record. An example about the concerns over the adversary's eventual forensic capabilities is layers of oil paint on a canvas; while previous layers appear to be deleted, X-ray technology has enabled the discovery of buried paintings, each appended over the previous [62].

3.4 Adversarial Model

We assume the presence of a computationally bounded, unpredictable, multiple access, coercive adversary. This means that the adversary gains access to the client's storage medium at multiple unpredictable points in time. This adversary can perform a *coercive attack* to compromise both the client's storage medium as well as any secret keys, etc., which may be needed to access data stored on the medium. The adversary has full knowledge of the algorithms as well as the implementation of the system and all relevant storage media. We make the assumption that symmetric-key cryptography is perfect, that is, the computationally bounded adversary cannot recover the plain-text message from a cipher-text message without the corresponding encryption key. This requires that any keys derived from passwords are taken from sufficiently strong passwords to prohibit offline guessing.

With the exception of the computational bound, this adversary is the strongest one developed in our taxonomy (cf. Table 2.1). Since the time of the attack is unpredictable, no extraordinary sanitization procedure can be performed prior to compromise. Since the user is continually using the storage media, physically destroying it is not possible. Since the attacker is given the user's secret keys, it is insufficient to simply encrypt the storage media [46]. The adversary may also attack multiple times if desired. The solutions we develop defeat this strong adversary and therefore all weaker adversaries as well.

3.5 Security Goal

Secure deletion is used to protect the confidentiality of data. For any particular data item, the security goal is to ensure that an adversary who compromises the storage medium at all times outside the data item's lifetime is unable to recover the data item. This is because we assume that an adversary that coercively attacks during some data's lifetime is easily able to recover the data; secure deletion aims to protect the confidentiality of data outside its lifetime. Observe that the security goal is realized for all data items if every time the adversary compromises the storage medium, it is *only* able to recover valid data; indeed, this is the behaviour of the optimal SECDEL model.

For clocked models, however, data is only deleted at the next clock edge. If an adversary compromises the storage medium after a data item is discarded but before the next clock edge, it obtains this discarded data. The time discarded data remains available is called the data item's *deletion latency*. We say that the security goal is achieved for all data items with a *deletion latency* of δ if every time the adversary compromises the storage medium it is *only* able to recover valid data or data discarded within the previous δ time units.

Paradoxically, it is also possible for data to be exposed to an adversary who compromises the storage medium *before* the data is written: this occurs when encryption keys used for data are written in advance of use (and, for example, the adversary trivially obtains encrypted data). Consequently, there is a period of time where adversarial compromises recover future data and we call this period the *existential latency*. We say that the security goal is achieved for all data items with a *existential latency* of ε if every time the adversary compromises the storage medium it is *only* able to recover valid data or data that is created within the next ε time units.

The behaviour of different models with regards to example adversarial compromises is illustrated with Figures 3.1–3.5. Each figure shows the storage history of seven data items—all with the same data lifetime—as stored by a storage medium that behaves like a particular model. The first, Figure 3.1, is for the optimal SECDEL model, while the last, Figure 3.5, is for the near-pessimal PERSISTENT model. Data lifetimes are visualized with a solid line with the create and discard events indicated with a black circle. A dotted line before or after these events indicate the existential and deletion latencies, respectively, that is, times when adversarial compromise obtains the data. The data effectively stored (i.e., compromisable) at each particular time is listed as a set below the life timelines.

Figure 3.1 shows the behaviour of the SECDEL model: only valid data is stored at each time. Figure 3.2 shows the behaviour of the SECDEL-CLOCK model with a clock period of four. Clock edges are shown as thick black lines and each data item has a deletion latency that extends to the next clock edge. Figure 3.3 shows the same for a clocked model that has both existential and deletion latencies, which we call SECDEL-CLOCK-EXIST. The data stored is the same for all times within a particular deletion epoch. Figure 3.4 shows the behaviour of either the INPLACE or SEMIPERSISTENT model. There are three different storage positions, and the dotted line that connects one data item to another indicates that some newly created data is stored in the same position as the previously deleted data. Figure 3.4 illustrates both INPLACE and SEMIPERSISTENT models because the main difference between them is whether the user can control the positions used to store new data. Finally, Figure 3.5 shows the behaviour of the PERSISTENT model, which stores all data that was earlier valid: the deletion latency extends to the end.

At the bottom of all these figures are four example adversarial compromises. Each adversary compromises the storage medium at a different set of times, indicated with a black box containing a time. The set of data obtained by each compromise is the union of data stored at each compromise time. The effectiveness of a model with regards to secure deletion can be seen by how well the adversary's *data obtained* approximates the SECDEL model.

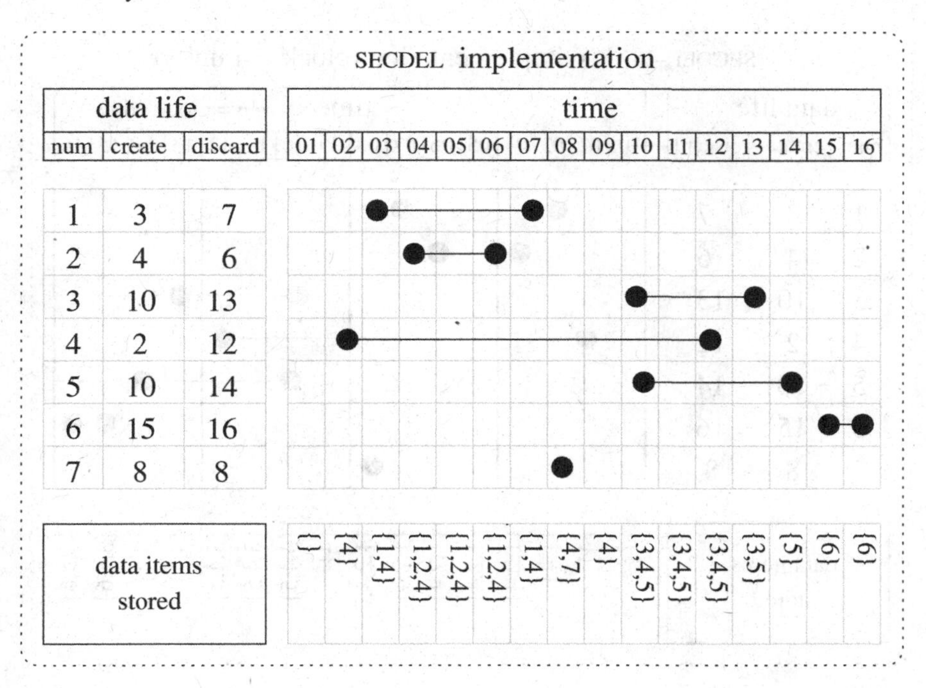

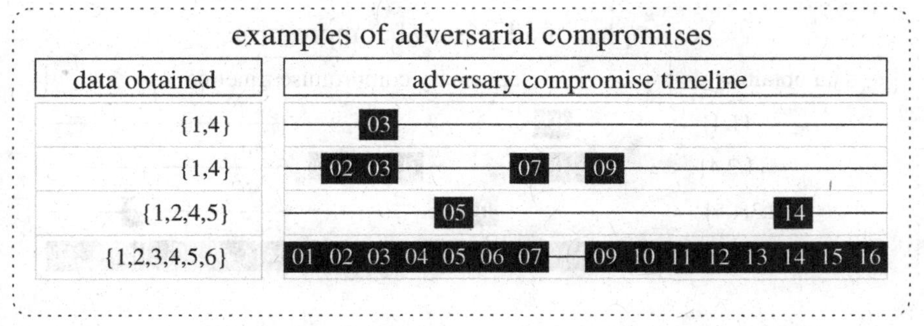

Fig. 3.1 Example data lifetimes and adversarial compromises for a SECDEL model. Seven numbered data items have each a lifetime, where time is discretized into 16 points in time. At each point in time, the set of valid data is indicated. Below the data lifetimes are four adversarial timelines. A black box with a number indicates that the adversary performs a coercive attack at that time. The adversary obtains all data valid at each time it compromises. We assume perfect secure deletion: the adversary only obtains valid data.

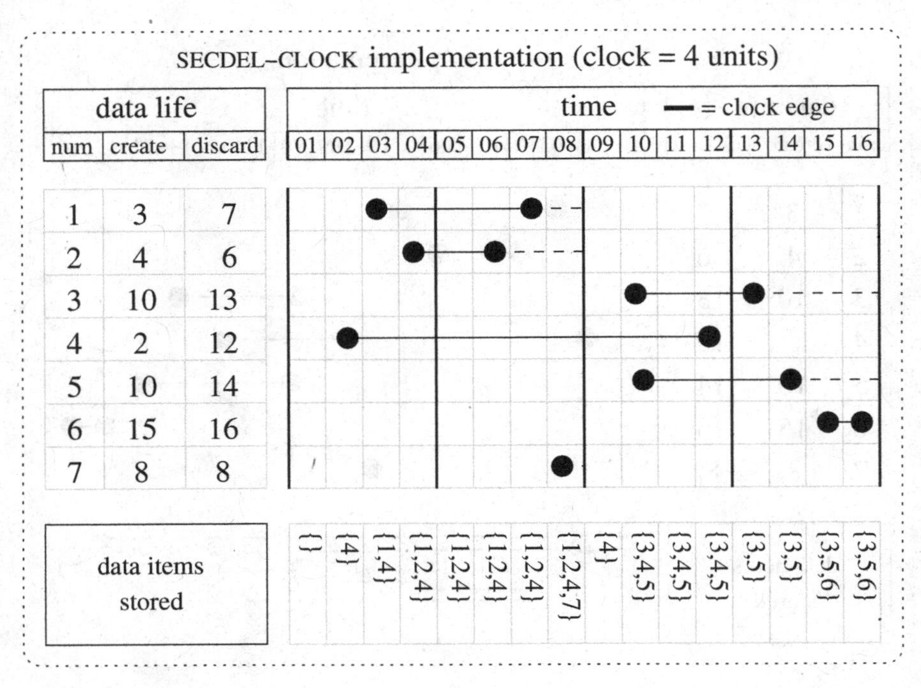

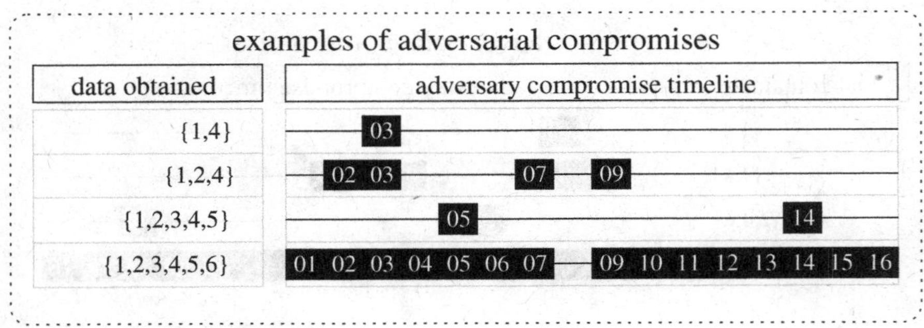

Fig. 3.2 Example data lifetimes and adversarial compromises for a SECDEL-CLOCK model. The data lifetimes and adversarial compromises are the same as Figure 3.1. We assume there is a clock operation performed every four time units, which is indicated with a thick black line in the data lifetimes. Data is only deleted at a clock operation, so a dotted line indicates the deletion latency. Adversarial compromises are updated accordingly to obtain deleted data which may still be available.

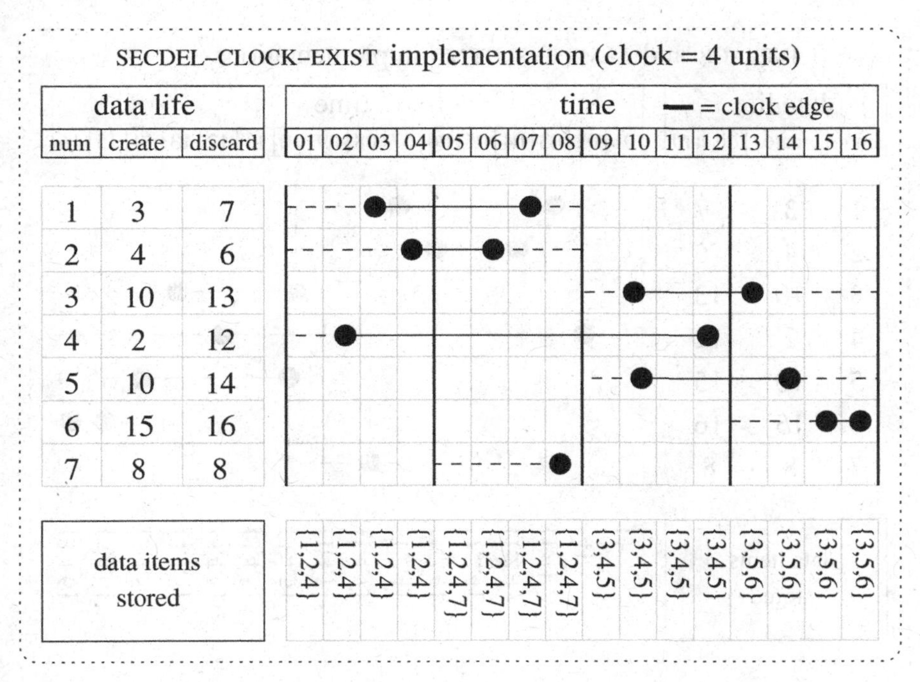

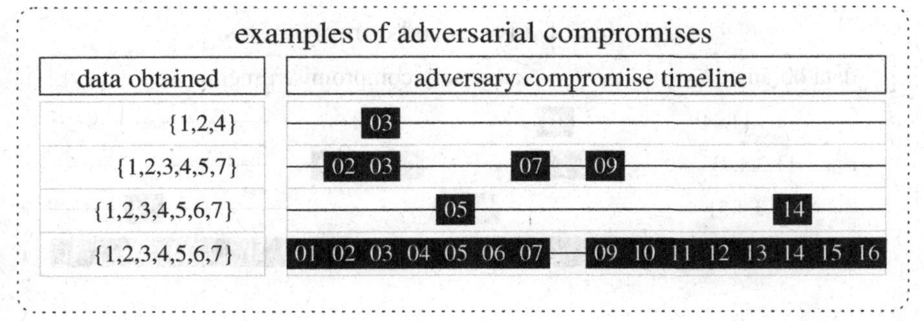

Fig. 3.3 Example data lifetimes and adversarial compromises for a SECDEL-CLOCK-EXIST model. The data lifetimes, adversarial compromises, and clock period are the same as Figure 3.2. Not only is data deleted at a clock edge, data can be obtained by compromising the medium before it is written provided it will be written in the same clock period. A dotted line before the data's creation indicates its existential latency. Adversarial compromises are updated accordingly.

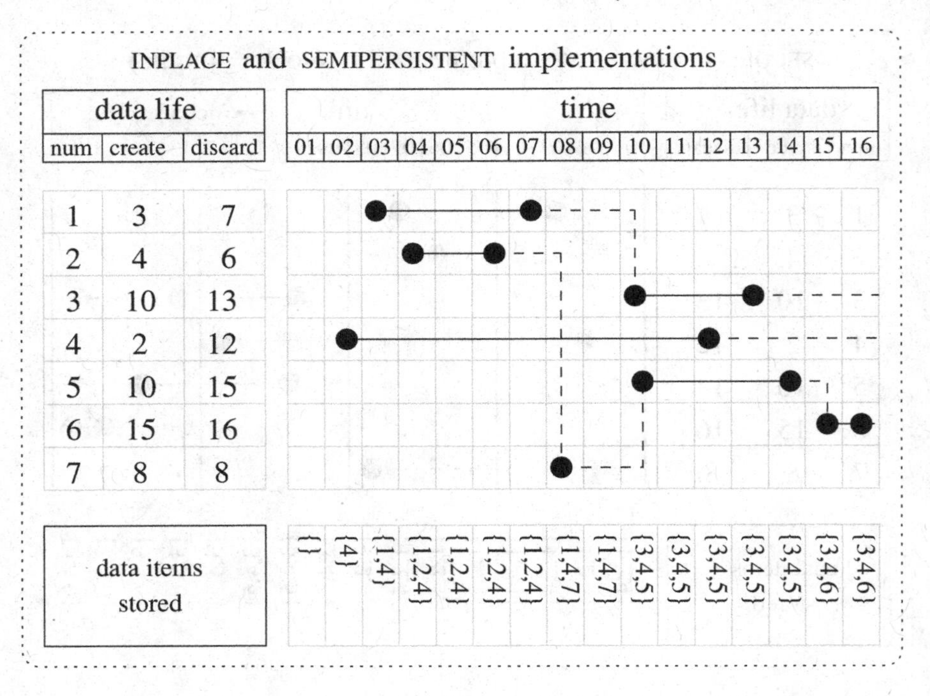

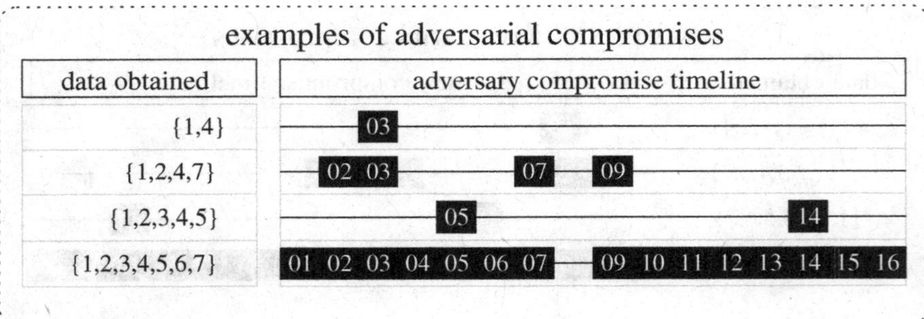

Fig. 3.4 Example data lifetimes and adversarial compromises for either an INPLACE or SEMIPERSISTENT model. There are three different "physical" storage positions; data remains stored until it is replaced with a new value. A dotted line from the deletion of some data to the creation of another piece of data indicates that the deleted data remained stored until that new piece of data is created. Adversarial compromises are updated accordingly. The difference between in-place updates and semipersistent updates is that in the former the user can select which data is replaced during an update.

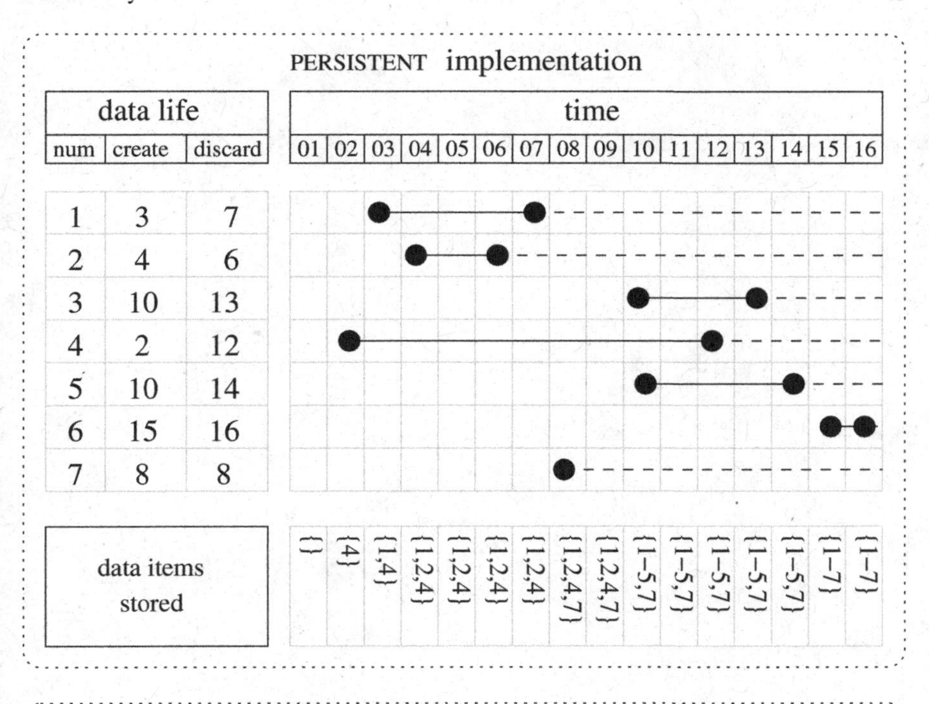

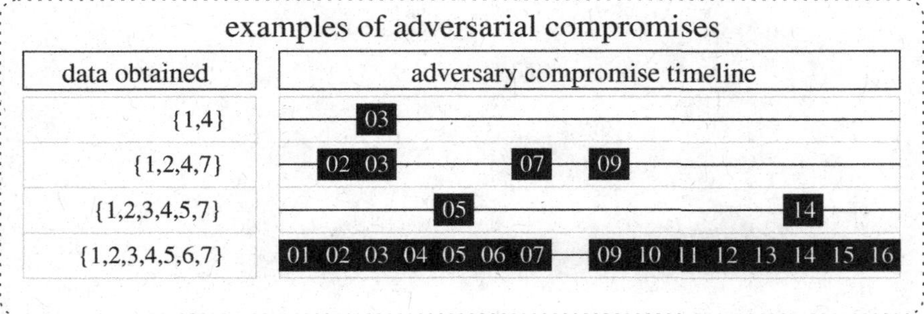

Fig. 3.5 Example data lifetimes and adversarial compromises for a PERSISTENT model. Once written, data remains stored permanently. Adversarial compromises are updated accordingly.

Part II
Secure Deletion for Mobile Storage

Chapter 4
Flash Memory: Background and Related Work

4.1 Overview

Flash memory is a fast, small, and lightweight type of storage medium that, at the time of writing, is ubiquitously used in mobile, portable devices, including mobile phones, MP3 players, digital cameras, voice recorders, gaming devices, USB sticks, multimedia cards, and solid-state drives. The ability to securely delete data from such devices is important because mobile phones, in particular, store sensitive personal information, such as the timestamped names of nearby wireless networks; personal correspondence; furthermore business data, for which company policy or legislation may mandate deletion after some time elapses or at some geographic locations.

Chapter 2 describes many secure deletion solutions suitable for magnetic storage; these solutions often involve overwriting the data to replace the old data on the medium. This overwriting technique does not work for flash memory, however, because flash memory cannot perform an in-place update—that is, an update that replaces the old version with a new version of data. Instead, all updates to flash memory are performed in a log-structured way: writing the fresh data to a new location and rendering the old version obsolete. In our model, flash memory has the behaviour of a SEMIPERSISTENT implementation.

In the following chapters, we present a detailed examination of secure-deletion solutions for flash memory. The remainder of this chapter presents details on flash memory including the interface to access it, flash file systems that use this interface, and flash translation layers that hide the details of flash memory and expose a block-based interface typical for magnetic media. We then present related work for secure deletion on flash memory.

Chapters 5, 6, and 7 provide our contributions. Chapter 5 presents our results on user-level secure deletion for flash memory, that is, what users can do to securely delete data on their mobile devices without changing their file system, operating system, or hardware. Chapter 6 presents the Data Node Encrypted File System (DNEFS), which augments a file system to provide secure deletion. Chapter 7 val-

J. Reardon, *Secure Data Deletion*, Information Security and Cryptography,
DOI 10.1007/978-3-319-28778-2_4

idates DNEFS by presenting an implementation for the flash file system UBIFS, analyzing its performance, and finding that it is suitable for secure data deletion for flash memory.

4.2 Flash Memory

Flash memory is a non-volatile storage medium consisting of an array of electronic components that store information [63]. A *page* of flash memory is *programmed* to store data, which can thereafter be *read* until the page is *erased* [64]. Flash memory has small mass and volume, does not incur seek penalties for random access, and is energy efficient. As such, portable devices ubiquitously use flash memory.

Figure 4.1 shows how flash memory is divided into two levels of granularity. The first level is called *erase blocks*, which are on the order of 128 KiB [65] in size. Erase blocks are divided into *pages*, which are on the order of 2 KiB in size. Note that different kinds of memory may have different sizes for the erase block and the page, however these values are representative of both typical memory devices as well as the difference in scale between the two levels of granularity.

Erase blocks are the unit of erasure, and pages are the unit of read and write operations [64]. One cannot write data to a flash memory page unless that page has been previously *erased*; only the erasure operation performed on an erase block prepares the pages it contains for writing.

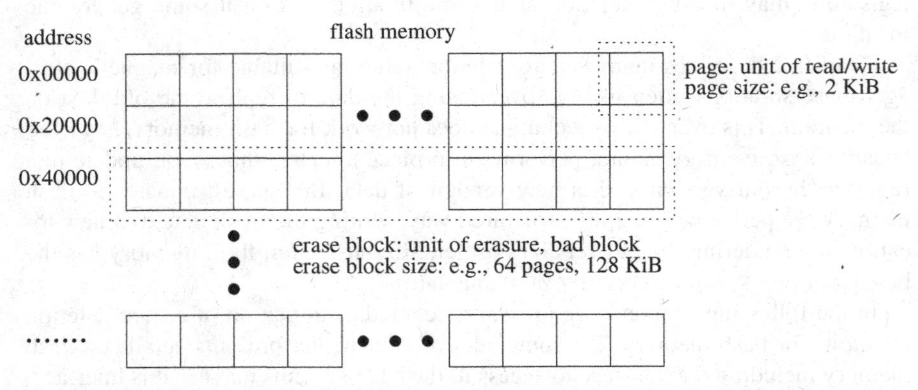

Fig. 4.1 Flash memory divided into two levels of granularity. Each row of squares represents an erase block and each square represents a page.

Flash erasure is costly: its increased voltage requirement eventually wears out the medium [66]. Each erasure risks turning an erase block into a bad block, which cannot store data. Flash erase blocks tolerate between 10^4 to 10^5 erasures before they become bad blocks. To promote a longer device lifetime, erasures should be evenly levelled over the erase blocks. This is commonly called *wear levelling*.

Flash memory best practices are that the erase block's empty pages are programmed sequentially by their physical layout. This mitigates an issue known as *program disturb*, where programming a flash page affects the data integrity of physically neighbouring pages. By programming pages sequentially, program disturb is only a concern for the most-recently programmed page.

In the remainder of this section, we describe how log-structured file systems are used to overcome flash memory's in-place update limitation. We then describe the two main types of log-structured implementations: (i) a flash translation layer that exposes a block device and (ii) special purpose flash file systems that expose POSIX-compliant file system interfaces.

4.2.1 In-Place Updates and Log-Structured File Systems

Flash memory's requirement that the entire erase block is erased before data can be written is the reason for flash memory's inability to perform in-place updates. Erase blocks may store a mix of deleted and valid data for a variety of different files. The naive way to update one page on an erase block is to temporarily store the entire erase block elsewhere, erase the original erase block, and finally reprogram all the pages on the original erase block to store the previous data with the exception of the updated page. In fact, Linux provides a simple block device emulator for flash memory, called `mtdblock` [67], which slightly improves on this naive update strategy by buffering multiple changes to a single erase block using a one-item memory cache.

In practice, flash memory's in-place update limitation is managed by using log-structured file systems to store and access data. A log-structured file system differs from a traditional block-based file system (such as FAT [68] or ext2 [44]) in that the entire file system is stored as a chronological record of changes from an initial empty state. As files are written, data is appended to the log indicating the resulting change; each flash page stores a fixed-size block of data. File metadata and data are usually stored without separation. The file system maintains in volatile memory the appropriate data structures to quickly find the newest version of each file header and data page [64, 69].

When a change invalidates an earlier change, then the new, valid data is appended and the erase block containing the invalidated data now contains wasted space. File deletion, for example, may append a log entry that states that file is thenceforth deleted. All the deleted file's data nodes remain on the storage medium but they are now invalid and wasting space. Encrypting a file, for example, appends a new encrypted version of that file with the obsolete plaintext now remaining in the log.

Data is removed from a log-structured file system with a *compaction* process often called *garbage collection* [64],[1] which has the purpose of reclaiming wasted storage resources. The compactor operates at the *erase block* level, which has a

[1] The term garbage collection comes from its similarity to garbage collection in programming languages where unused memory is found and collected; in this book we use the term *compaction*

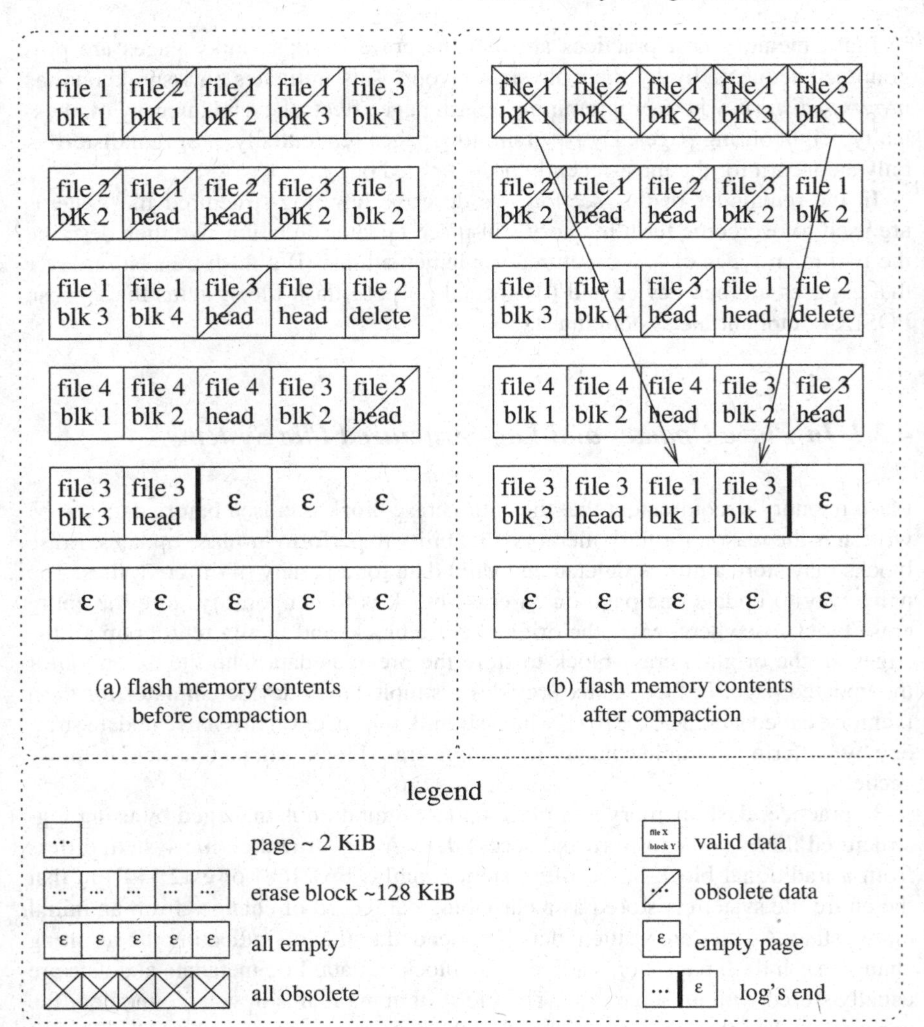

Fig. 4.2 Flash memory storing a log-structured file system. Data blocks and headers for different files are stored on the pages; some pages are obsolete as newer versions exist. File 2 is at one point deleted, making all file data except the delete notice obsolete. (a) The state before compacting the first erase block. (b) Two valid pages are copied from the first erase block to the end of the log. The first erase block can now be erased.

larger granularity than a page. While implementation details may vary, in principle the compactor erases any erase block that only contains deleted data, and also compacts erase blocks with a significant amount of wasted space by first copying live data to the log's end and then erasing the old erase block. Figure 4.2 shows a flash

to refer to the "garbage collection" process that specifically collects *non-garbage* data on erase blocks to recover wasted resources.

memory being used to store a log-structured file system's data and the compaction of the first erase block. Figure 4.2 (a) shows the state before compaction where the first erase block contains only two valid blocks; (b) shows the state after copying the valid data to the log's end, resulting in the first erase block storing only obsolete data. The erasure operation can then be performed on it to recover the wasted resources.

As a historical observation, log-structured file systems were first implemented in 1992 by Rosenblum and Ousterhout without anticipating flash memory. Their purpose was to improve efficiency for file systems that perform frequent small writes to different files—i.e., file systems used for logging events—by putting all new data at the end of the log to reduce magnetic media's high seek latency [70]. The notion of erase blocks, then called *segments*, was needed for such systems because fragmentation in the sequence of writable positions removed the benefit of seek-free writes. The *segment cleaner* therefore compacted the useful data elsewhere and reclaimed the entire segment for fresh data. This idea ultimately found enormous utility in mitigating the quirks of flash memory—a memory that does not even suffer from seek latency.

4.2.2 Flash Translation Layer

Flash memory is commonly accessed through a Flash Translation Layer (FTL) [63, 71], which is used in USB sticks, MMC devices such as SD cards, and solid state drives. FTLs access the raw flash memory directly, but expose a block-based interface that is typical for magnetic hard drives. This allows users to use widely compatible file systems—in particular, FAT—when storing data, allowing easy exchange of data across devices and computers.

FTLs vary in implementation [72, 73], however they all have a simple purpose: to translate logical block addresses to raw physical flash addresses and internally implement a log-structured file system on the memory [72]. New data is written to empty flash memory pages (the end of the log) and a mapping keeps track of what is the most recent version of a particular logical address in the virtual block device.

Most FTLs used for portable devices are implemented in hardware. In this case, software access to the device can only reveal the contents of the virtual block device. For example, the specification for embedded multimedia cards (eMMC) provides no interface functionality to read data from *physically unmapped areas*, that is, parts of the physical memory that do not correspond to data *officially* stored on the virtual block device. By disassembling the hardware and accessing the flash memory directly, however, one can bypass the hardware FTL and easily read the stored data [26]. Linux offers an `ftl` driver as an open-source software-based implementation of an FTL based on Intel's specification [71].

4.2.3 Flash File Systems

Another solution to accessing flash memory is to use a file system tailored for the purpose. A variety of flash file systems (FFSs) exist; open-source ones for Linux include JFFS [74], YAFFS [69], UBIFS [75], and F2FS [76]. These file systems are log-structured and access the flash memory directly through a Memory Technology Device (MTD). MTDs offer a similar interface as block devices do but are extended to support erasing an erase block, testing if an erase block is bad (i.e., unwritable), and marking an erase block as bad.

Using a tailored flash file system means that a magnetic-storage-targeted file system—which may replicate features such as journalling—need not be mounted above an FTL. Moreover, the lack of an opaque hardware FTL permits greater transparency in how data is stored and deleted as well as easier integration and verification of secure deletion solutions. These file systems, however, are not widely supported outside Linux systems and therefore are less suitable as external (removable) memory than as internal (non-removable) memory.

4.2.4 Generalizations to Other Media

The asymmetry between the write and erase granularities is not limited to flash memory, and Table 4.1 summarizes such storage media. It manifests itself in physical media composed of many write-once read-many units; units that are unerasable but replaceable. Examples include a library of write-once optical discs or a stack of punched cards. All write-once media are unerasable—NIST says they must be physically destroyed to achieve any form of secure deletion [11]—but first valid colocated data must be replicated onto a new disc or card and then the library updated. Therefore, each erase operation performed on such media destroys one of its constituent storage units.

Similarly, media that can be erased but only with a high asymmetry in granularity also suffer from this problem. For example, a tape archive consists of many magnetic tapes, each storing, say, half a terabyte of data. Tape must be written end-to-end in one operation; data available for archiving is heuristically bundled onto a tape. Later, to securely delete a single backup on the tape, the entire tape is re-written to a new tape with the backup removed or replaced; the old tape is then erased and reused in the tape archive. This operation incurs cost: tapes have a limited erasure lifetime and tape-drive time is an expensive resource for highly utilized archives.

4.3 Related Work for Flash Secure Deletion

In this section, we describe related work on the topic of secure deletion for flash memory.

Table 4.1 Generalizations of the write and erase granularity asymmetry to other storage media.

media	collection	I/O unit	erase unit	erase op.	relevant cost
optical disc	library	track	disc	destroy	blank media
magnetic tape	vault	backup	cassette	tape-over	tape wear, drive time
flash memory	memory	page	erase block	erasure	erase block wear, power
punched cards	stack	column	card	shred	blank media, repunching

Secure Erase / Factory Reset.

Some flash-based devices offer a *factory reset* feature, which acts like a secure erase feature in their hardware controllers. Such a feature is intended to perform erase block erasure on all the erase blocks that comprise the storage medium. This means that the solution has a per-storage-medium granularity.

A study by Wei et al. [26] observed that solid-state drives' controller-based *secure erase* operation is occasionally incorrectly implemented. In some cases, the device reported a successful operation while the entire file system remained available. In follow-up work, Swanson and Wei [7] describe a solution for verifiable full-device secure deletion that they compare in effectiveness to hard drive degaussing. They propose to encrypt all data written to the physical medium with a key stored only on the hardware controller. To securely delete the device, first the controller's key memory is erased. Every block on the device is then erased, written with a known pattern, and erased again. Finally, the device is reinitialized and a new key given to the flash controller. As with the factory reset, this solution has a per-storage-medium granularity.

The eMMC specification states that the *secure erase* and *secure trim* functions are obsolete as of v4.51 [61]. Instead, a *sanitize* (i.e., secure deletion) function may be optionally provided. The specification states that the sanitize function must erase all erase blocks containing unmapped data including all copies; mapped data must remain available. If correctly implemented, this provides secure deletion at a per-data-item granularity; verifying that it is correctly implemented, however, requires disassembling the physical device. The Linux eMMC device allows this operation to be performed from user-space via an `ioctl` [34].

Compaction.

The naive secure-deletion solution for physical media with an asymmetry between their write and erase granularities is to immediately compact the erase block that contains the deleted data: copy the valid colocated data elsewhere and execute the erasure operation. This is a costly operation: copying the data costs time, and erasing an erase block may additionally cause wear on the physical medium. Nevertheless, there is no other *immediate* secure-deletion solution based on erasures that can do better than one erase block erasure per deletion. Any improvement requires batching and thus effects a deletion latency.

Batched Compaction.

One obvious improvement over the naive solution is to intermittently perform compaction-based secure deletion on all the erase blocks that have accumulated deleted data since the last secure deletion. This solution is no worse than the naive solution in terms of the time and wear, although the deletion latency—the time the user must wait until data is securely deleted—increases. Each time that deleted data items are colocated on an erase block, the amortized time and wear cost of secure deletion decreases. Indeed, log-structured file systems already perform a similar technique to recover wasted space, where compaction is performed only on erase blocks whose wasted space exceeds a heuristically computed threshold based on the file system's current need for free space.

Per-File Secure Deletion.

Lee et al. [77] propose a secure deletion solution for YAFFS [69]. It performs immediate secure deletion of an entire file at the fixed cost of one erase block compaction. It reduces the erasure cost of secure deletion by only deleting at a per-file granularity: Until the file is deleted, it remains entirely available including overwritten and truncated parts. When the file is deleted, a single erase block erasure is sufficient to ensure it is securely deleted.

Their solution encrypts each file with a unique key stored in every version of the file's header. The file system is modified to store all versions of a file's header on the same erase block. Whenever erase blocks storing headers are full, they are compacted to ensure that file encryption keys are only stored on one erase block. To delete a file, the erase block storing the key is compacted for secure deletion, thus deleting all file data under computational assumptions with only one erase block erasure.

Scrubbing.

Compaction is the only immediate secure deletion solution that uses erasures. Wei et al. [26] propose a solution for flash memory, called *scrubbing*, which works by draining the electrical charge from flash memory cells—effectively rewriting the memory to contain only zeros.

Scrubbing securely deletes data immediately with the granularity of a page and no erase block erasures must be performed. It does, however, require programming a page multiple times between erasures, which is not appropriate for flash memory [66]. In general, the pages of an erase block must be programmed sequentially [78] and only once. Multiple partial programs per page is permitted provided that they occur at different positions in the page and are fewer than the manufacturer's specified limit; multiple overwrites to the same location officially result in undefined behaviour [78]. Flash manufacturers prohibit this due to *program disturb* [66]: bit errors that can be caused in spatially proximate pages while programming flash memory.

Wei et al. performed experiments to quantify the rate at which such errors occur; they showed that they do exist but their frequency varies widely among flash types, a result also confirmed by Grupp et al. [79]. Wei et al. use the term *scrub budget* to refer to the number of times that the particular model of flash memory has experimentally allowed multiple overwrites without exhibiting a significant risk of data errors. When the scrub budget for an erase block is exceeded, then secure deletion is instead performed by compaction: copying all the remaining valid data blocks elsewhere and erasing the block. Wei et al.'s results show that modern densely packed flash memories are unsuitable for their technique as they allow as few as two scrubs per erase block [26].

4.4 Summary

While efforts have been made to solve the problem of secure deletion for flash memory, none of these solutions is perfect. Factory reset suffers from problems associated with per-storage-medium solutions. Compaction deletes the data but comes at a high cost in terms of flash memory erasures. Lee et al.'s per-file encryption suffers from problems associated with per-file solutions and reduces to naive compaction when dealing with many small files. Scrubbing does not work with all flash memory and, in particular, works less effectively on newer devices. We therefore need novel solutions to this problem.

In the next two chapters, we present our efforts to solve this problem. Chapter 5 presents user-level secure deletion for flash memory. It provides two solutions, as well as a hybrid of them, which provide secure deletion functionality for users without having to modify their operating system or file system. We perform experiments to measure the solutions' costs in terms of erase block erasures and their benefits in terms of deletion latency.

Chapter 6 presents the Data Node Encrypted File System. It is a generic extension to a file system, and therefore a kernel-level solution, which can be used to provide secure data deletion with a configurable deletion latency. It significantly reduces the number of erase block erasures required to achieve secure deletion.

Chapter 7 validates our file system extension by implementing it for the flash file system UBIFS. We measure its performance both in simulation and deployed on a mobile phone and show that it results in a modest increase in the erase block erasure rate and modest decrease in performance.

Chapter 5
User-Level Secure Deletion on Log-Structured File Systems

5.1 Introduction

This chapter addresses the problem of secure data deletion on log-structured file systems. We focus on YAFFS, a file system used on Android smartphones that uses raw flash for the internal memory. We analyze how deletion is performed in YAFFS and show that log-structured file systems in general provide no temporal guarantees on data deletion; the time discarded data persists on a log-structured file system is proportional to the size of the storage medium and related to the writing behaviour of the device using the storage medium. Moreover, discarded data remains stored indefinitely if the storage medium is not used after the data is marked for deletion.

We propose two user-level solutions for secure deletion in log-structured file systems: *purging*, which provides guaranteed time-bounded deletion of all previously discarded data by filling the storage medium, and *ballooning*, which continuously reduces the expected time that any discarded data remains on the medium by occupying a fraction of the capacity. We combine these two solutions into a hybrid, which guarantees the periodic, prompt secure deletion of data regardless of the storage medium's size and with acceptable wear of the memory.

As these solutions require only user-level permissions, they enable the user to securely delete data even if this feature is not supported by the kernel or hardware, over which users typically do not have control. This, for example, allows mobile phone users to achieve secure deletion without violating their warranties or requiring non-trivial technical knowledge to update their firmware with a customized kernel. Nevertheless, user-level solutions have a reduced interface to the storage medium and so cannot be more efficient than solutions with a deeper integration; in the next chapter we see exactly such a solution that achieves greater efficiency.

We implement these solutions on an Android smartphone (Nexus One [65]) and show that they neither prohibitively reduce the longevity of the flash memory nor noticeably reduce the device's battery lifetime. We simulate our solutions for phones with larger storage capacities than the Nexus One, and show that while purging alone is expensive in time and flash memory wear, when combined with ballooning

J. Reardon, *Secure Data Deletion*, Information Security and Cryptography,
DOI 10.1007/978-3-319-28778-2_5

it becomes feasible and effective. Ballooning provides a trade-off between the deletion latency and the resulting wear on the flash memory. It also substantially reduces the deletion latency on large, sparsely occupied storage media.

5.2 System and Adversarial Model

The user continually stores, reads, and discards sensitive data on a mobile phone. We assume that the user has only user-level access to the mobile phone. This means that the user may not modify the operating system or hardware of the device. The solution can only interact with the file system interface to achieve secure deletion.

We assume that there is an unpredictable multiple-access coercive adversary that can compromise the user's storage medium. Our adversarial model is a slight modification of the main model developed in Chapter 3 in that it is not computationally bounded.

5.3 YAFFS

Yet Another Flash File System (YAFFS) is a log-structured file system designed specifically for flash memory [69]. It is notably used as the file system for the internal memory of some Android mobile phones which store data using raw flash memory.

YAFFS allocates memory by selecting an unused erase block and sequentially allocating the numbered pages (which YAFFS calls chunks) in that erase block. An allocated erase block is freshly erased and therefore devoid of any data. YAFFS searches for empty erase blocks (i.e., ones that contain no valid data) sequentially by the erase block number as defined by the physical layout of memory on the storage medium, wrapping cyclically when necessary. It begins searching from the most-recently allocated erase block and returns the first empty erase block.

YAFFS performs *compaction* (which YAFFS calls garbage collection) to reclaim wasted space on partially full erase blocks. As illustrated in Figure 4.2, compaction copies all valid (i.e., non-discarded) pages from some partially full erase block to the log's end; compaction then erases the source erase block, which now contains no valid data. If there is no erase block that can be compacted, that is, there is not a single unneeded page stored on the medium, then YAFFS reports the file system as full and fails to allocate an erase block.

Compaction in YAFFS is either initiated by a thread that performs system maintenance, or takes place during write operations. Usually, a few pages are copied at a time, thus the work to copy an erase block is amortized over many write operations. If the file system contains too few free erase blocks, then a more aggressive compaction is performed. In this case, erase blocks with *any* amount of discarded space are compacted.

YAFFS selects erase blocks for compaction using a greedy strategy based on the ratio of discarded pages on an erase block; however, it only searches within a small moving range of erase blocks with a minimum threshold for discarded pages. This cyclic and proactive approach to compaction results in a strong cyclic trend in erase block allocations. When low on free space, YAFFS selects the erase block with the most wasted space by examining all the storage medium's erase blocks.

There are currently two major versions of YAFFS, YAFFS1 and YAFFS2, and among their differences is how file deletion is performed. In YAFFS1, a special *not-deleted* flag in the file's header is set to **1**; when the file is deleted the header is programmed a second time (without first erasing it) to contain the same contents except the flag is set to **0**. Note that this technique is similar to Wei et al.'s *scrubbing* [26]. In YAFFS2, this multiple programming is obviated by writing a new file header instead; this change is to allow YAFFS to support all flash memories, many of which do not permit multiple programmings. We used YAFFS2 for all our experiments and henceforth we simply call it YAFFS.

5.4 Data Deletion in Existing Log-Structured File Systems

In this section, we investigate data persistence on log-structured file systems by analyzing the internal memory of a Nexus One running Android/YAFFS and simulating larger storage media. We instrument the file system at the kernel level to log erase block allocation information. This provides an upper bound on the deletion latency, because allocating an erase block for storage implies that it was previously compacted and erased, and therefore all discarded data previously stored *on that erase block* is securely deleted.

Figure 5.1 shows four data items stored on two erase blocks, each one with two pages. Different data items are indicated with different patterns. Each erase block shows a timeline of what data is stored on which page at which time. Data item create and discard events are indicated. Moreover, when an erase block is reallocated, valid data is copied to another page. The vertically striped data item, for example, is twice copied before it is discarded. At the bottom is illustrated the lifetime of data items as well as the time that they are compromisable due to deletion latency. The time between two erase block reallocations is labelled as the erase block reallocation period. All data discarded within this period has its deletion latency bounded by it. Observe that the reallocation period is not fixed for all times and erase blocks; it depends on how the storage medium is used.

In this section we show that modern Android smartphones have large deletion latency, where deleted data can remain indefinitely on the storage media. This motivates the secure deletion solutions in Section 5.5.

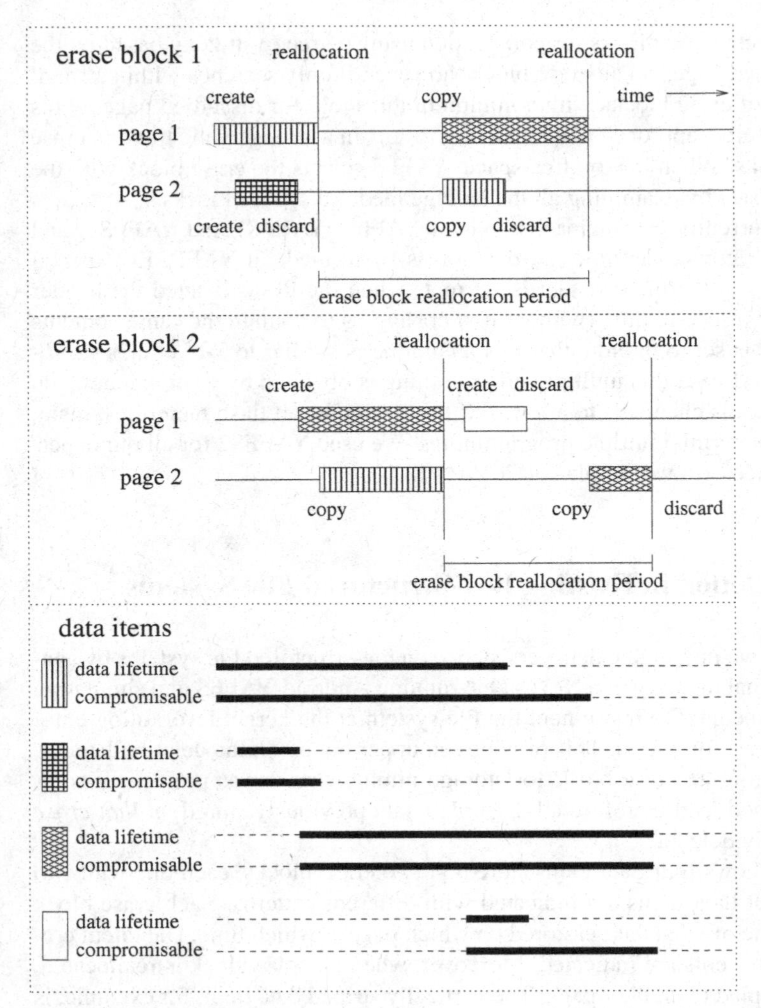

Fig. 5.1 Example timeline of data items stored on two erase blocks. Each erase block is twice reallocated and the reallocation period for them is indicated. Different data items have different patterns. The bottom illustrates each data item's lifetime and compromisable time (i.e., lifetime plus deletion latency).

5.4.1 Instrumented YAFFS

We built a modified version of the YAFFS Linux kernel module that logs data about the writing behaviour of an Android phone. We log the time and number for every erase block allocation and erasure. This information shows us where YAFFS stores data written at some point in time and when that data becomes irrecoverable. This allows us to compute the deletion latency of data in our simulation.

We used the instrumented phone daily for 670 hours, roughly 27.9 days. Throughout the experiment we recorded 20345 erase block allocations initiated by 73 different *writers*. A *writer* is any application, including the Android OS itself or one of its services (e.g., GPS, DHCP, compass, etc.). The experiment's logs show that the median time between erase block reallocations is 44.5 hours. The deletion latency is always less than the reallocation period; this means that the median deletion latency is upper bounded by this value.

5.4.2 Simulating Larger Storage Media

Log-structured file systems favour allocating empty erase blocks before compacting partially empty erase blocks [69, 73]. We hypothesize that the erase block reallocation period—and consequently the deletion latency—is highly dependent on the file system's size. We tested this hypothesis by writing a discrete event simulator to experiment with the writing behaviour of an Android phone on simulated YAFFS storage media of various sizes. We first describe our experimental setup and then present our results.

Experimental Procedure.

To experiment with different flash storage medium sizes, we simulated an Android mobile phone using a flash storage medium in memory. We used our own discrete event simulator that writes, overwrites, and deletes files on a storage medium. This medium is a directory on our computer that simulates accessing flash memory through a flash file system.

We used the collected statistics from our instrumented phone in Section 5.4.1 to determine the writing behaviour for our discrete event simulator. We logged every page that was written to the device for a week, and used this data to compute the period between successive creations of new files, and the characteristics of the files that are created. The characteristics of files are the following:

- The file's lifetime.
- A distribution over the period of time between opening a file for write.
- A distribution over the number of pages to write to a file each time it is opened.
- A distribution over a file's pages where the writes occur.

Additionally, we implemented a pattern writer that operated alongside the simulated writers. It periodically writes a one-page pattern, waits until a new erase block is allocated, and then deletes the pattern. We use the pattern writer to determine the deletion latency for data written at that particular moment in time, but which remains stored; it represents the writing of some sensitive data that is later discarded.

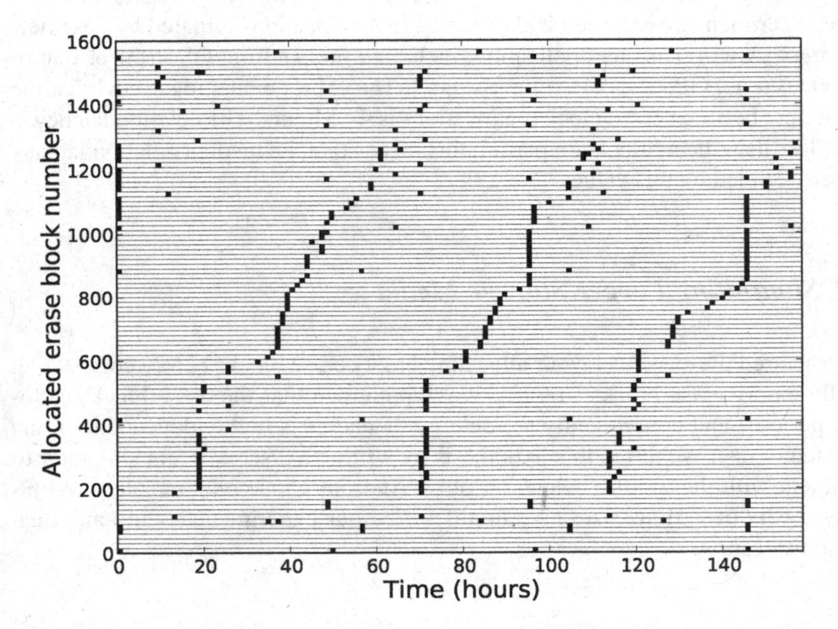

Fig. 5.2 Sampled plot of erase block allocation over time for YAFFS on an Android phone. The time between two points on the same horizontal line is the erase block reallocation period.

We perform experiments using YAFFS mounted on a virtual flash storage medium created by the kernel module `nandsim`. We use an erase block size of 64 2-KiB pages, consistent with the Nexus One phone [65].

Deletion Latency.

Figure 5.2 shows a plot of the storage medium's erase block allocations over time to gain insight on its behaviour. The horizontal axis is time, and the vertical axis shows the sampled space of sequentially numbered erase blocks. A black square on the graph means that an erase block was allocated at that time. For clarity, we compress the space of erase blocks into the rows by sampling every 15th erase block.

We present the results of our experiment in Table 5.1, which gives the median and 95th percentile deletion times in hours for the patterns written onto the storage medium during simulation. The maximum deletion latency is undefined because these systems provide no deletion guarantee and some data remained available after the experiment. Table 5.1 provides results for YAFFS partitions with sizes 200 MiB, 1 GiB and 2 GiB based on our observed access patterns.

We observe the effect of cyclic erase block allocation in YAFFS. There is both a linear growth in deletion latency as the size of the partition increases, and a high

Table 5.1 Deletion latency in hours for different configuration parameters.

partition size and type	deletion latency (hours)	
	median	95th percentile
200 MiB YAFFS	41.5 ± 2.6	46.2 ± 0.5
1 GiB YAFFS	163.1 ± 7.1	169.7 ± 7.8
2 GiB YAFFS	349.4 ± 11.2	370.3 ± 5.9

percentile observation close to the median. For instance, a YAFFS implementation on a 2 GiB partition (e.g., the data partition on the Samsung Galaxy S [80]) with the same access patterns can expect deleted data to remain up to a median of two weeks before actually being erased. In the next section, we present solutions to reduce this data deletion latency.

5.5 User-Space Secure Deletion

In this section, we introduce our solutions for secure deletion: purging, ballooning, and a hybrid of both. These solutions all work at user level, which has a limited interface that can only create, modify, and delete the user's own local files. Such solutions cannot force the file system to perform erase block erasures, prioritize compaction of particular areas in memory, or even know where on the storage medium the user's data is stored.

All of the solutions we present operate with the following principle: they reduce the file system's available free space to encourage more-frequent compaction, thereby decreasing the deletion latency for deleted data. Purging consists of filling the storage medium to capacity, thus ensuring that no deleted data can remain on the storage medium. Purging executes intermittently and halts after completion. Ballooning continually occupies some fraction of the storage medium's empty space to ensure it remains below a target threshold, thereby reducing the deletion latency. Ballooning executes continually during the lifetime of the storage medium. The hybrid solution performs ballooning continually, and performs a clock-driven purge operation to guarantee an upper bound on deletion latency.

We implement our solutions and examine their effectiveness for various storage medium sizes. We use deletion latency and storage medium wear as metrics for evaluating their effectiveness. We show that the hybrid solution is well suited for large storage media, where the deletion latency is a trade-off with storage medium wear.

5.5.1 Purging

Purging attempts to completely fill the file system's empty space with junk files; if the operation is successful then all partially filled erase blocks on the storage medium are compacted, and therefore all previously discarded data is securely deleted. Importantly, whether completely filling the file system from user space actually completely fills the storage medium depends on the implementation of the actual file system.

After filling the storage medium, the junk files are deleted so that the file system can again store data. Purging must be explicitly executed, which can take the form of automated triggers: when the phone is idle, when the browser cache is cleared, or when particular applications are closed. It is also useful for employees who are contractually obligated to delete customer data, e.g., before crossing a border.

The fact that the storage medium must be completely filled follows from a worst-case analysis of a SEMIPERSISTENT implementation whose allocatable space is the same as the addressable space. Before the storage medium is completely full, there is some area of the medium containing one last piece of unneeded but available data—we must pessimistically assume that is our discarded data. It is important to note that purging's ability to securely delete data is dependent on the implementation of the log-structured file system. In particular, we require the following condition to hold: if the file system reports that it is out of space, then all previously deleted pages are no longer available on the storage medium. While this condition holds for YAFFS, the implementation of other flash file systems and FTL hardware may differ.

A natural concern for purging's correctness is its behaviour on multithreaded systems. However, using the previous reasoning, purging needs to keep writing to the storage medium until it reports that it is completely full. This ensures that any data that has been deleted prior to purging is irrecoverable as the drive is completely full. Another concern is that, at the moment the storage medium is full, other applications simultaneously writing to the storage medium are told that the storage medium is full. We observe that any ungraceful handling of an unwritable storage medium is a flaw in the application and the storage medium's lack of capacity is a temporary condition that is quickly relieved.

We tested purging with the following experiment. We took a pristine memory snapshot of the phone's internal NAND memory by logging into the phone as root, unmounting the flash storage medium, and copying the raw data using `cat` from `/dev/mtd/mtd5` (the device that corresponds to the phone's data partition) to the phone's external memory (SD card). We wrote an arbitrary pattern not yet written on the storage medium, and obtained a memory snapshot to confirm its presence. We then deleted the pattern, obtained a new memory snapshot, and confirmed that the pattern still remained on the flash memory. Finally, we filled the file system to capacity with a junk file, deleted it, and obtained another memory snapshot to confirm that the pattern was no longer on the flash memory.

The time it took to execute purging on the Nexus One was between thirty seconds and a minute. As we soon see, however, this time is highly dependent on the storage

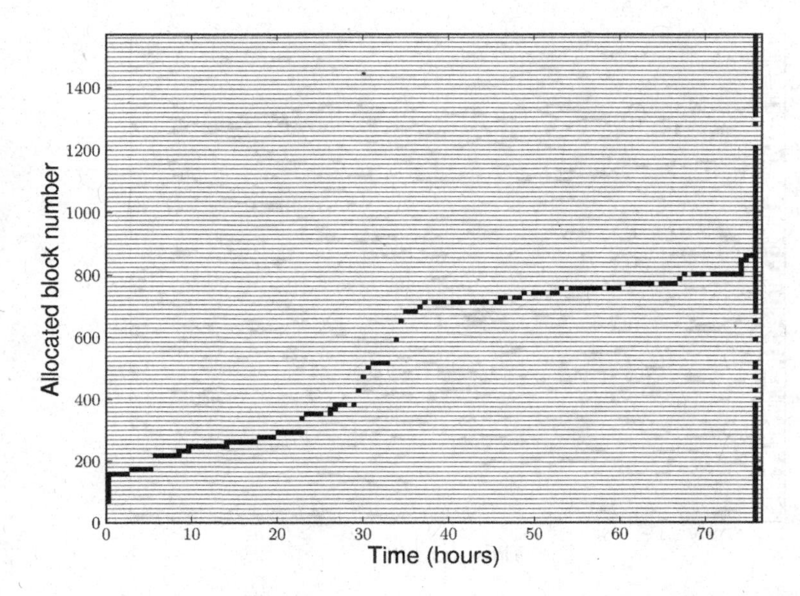

Fig. 5.3 Plot of erase block allocation over time for YAFFS (cf. Figure 5.2). After simulating writing for some time, we performed purging, which is visible at the right edges of the plot where many erase blocks are rapidly allocated.

medium's size. During execution the system displayed a warning message that it was nearing drive capacity, but the warning disappeared after completion.

Figure 5.3 shows the resulting erase block allocations reported by an instrumented version of YAFFS executing purging. The horizontal axis corresponds to time in hours, and the vertical axis shows the sampled space of numbered erase blocks. A small black square in the graph indicates when each erase block was allocated. For clarity, as with Figure 5.2, only a sampled subset of erase blocks (every 15th) have their allocations plotted. At the right side of Figure 5.3, we see the near immediate allocation of every erase block on the medium as indicated by the black squares forming a near vertical line. This is the consequence of filling the storage medium to capacity; a log-structured file system must compact every erase block that contains at least one deleted page.

5.5.2 Ballooning

In contrast to purging, which guarantees secure data deletion with a bounded deletion latency, we now present ballooning, which does not guarantee secure deletion with any bound but does reduce the deletion latency in expectation. Bal-

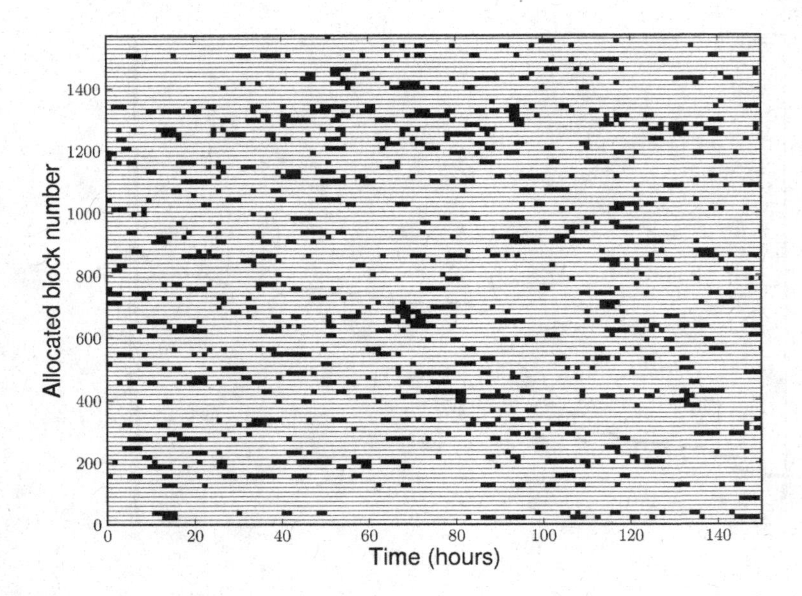

Fig. 5.4 Plot of erase block allocation over time for YAFFS while using aggressive ballooning.

looning artificially constrains the file system's available free space. This results in more-frequent compaction due to reduced capacity, and therefore reduces the time any deleted data—regardless of *when* it is deleted—remains accessible on a log-structured file system. Ballooning creates junk files to occupy the free space of the storage medium, which reduces the total number of erase blocks available for allocation. This reduces the expected erase block reallocation period, and therefore the expected deletion latency. These ballooning junk files are periodically rotated—new ones written and then old ones deleted—to promote efficient wear levelling.

In Section 5.6, we explore how varying free space thresholds—the aggressiveness of ballooning—affect deletion latency and other measurements. First, however, we visualize evidence that does not refute our hypothesis that ballooning reduces the erase block reallocation period. Figure 5.4 shows the erase block allocations that result from executing ballooning on YAFFS. We see a stark difference when compared with Figure 5.2. As the number of allocatable erase blocks decreases, YAFFS' sequential allocation becomes much more erratic, and the erase block reallocation period decreases. Row segments in Figure 5.4 that contain no allocation activity (i.e., a black square) likely correspond to erase blocks that are now filled with junk files. The figure shows a decrease in the erase block allocation period, which therefore reduces the expected deletion latency.

5.5.3 Hybrid Solution: Ballooning with Purging

The disadvantage of purging is that its cost is dependent on the free space available on the storage medium. In contrast, the disadvantage of ballooning is that it cannot provide a guarantee on when (or indeed if) data is deleted. By combining both these solutions, we create a hybrid scheme that has neither disadvantage. We use periodic purging for secure data deletion, and we use ballooning to ensure that a large storage medium's empty space must not be refilled during every purging operation. The result is a clock-based solution where purging is periodically performed, dividing time into deletion epochs. The deletion latency of all data is therefore bounded by the duration of a deletion epoch. The resulting storage medium has a SECDELCLOCK behaviour.

Reducing the number of erase blocks that must be filled during purging mitigates three concerns: purging's wear on the storage medium, its power consumption, and its execution time. Large capacity storage media are particular suitable to this solution: they may have large segments of their capacity empty, which ballooning occupies with junk files to achieve a deletion latency representative of smaller-sized storage media. In the next section we quantify this with experimental results for various storage medium sizes and ballooning aggressiveness settings.

5.6 Experimental Evaluation

We developed an application that implements our hybrid solution. The application periodically examines the file system to determine the free space, and appropriately creates and deletes junk files to maintain the free space within the upper and lower thresholds. The lower threshold is user defined and we set the upper threshold to be 4 MiB larger than the lower threshold to avoid a thrashing effect. The oldest junk file is always deleted before more recent ones to load-balance flash memory wear. Long-lived junk files can also be removed, with new ones written, to perform appropriate wear-levelling if necessary. The purging interval is user specified, allowing the user to select a tradeoff between the timeliness of secure deletion and the resulting wear on the device.

Our application runs successfully on the Android phone. The only permission it requires is the ability to run while the phone is in a locked state; the application also needs to specify that it runs as a service, meaning execution occurs even when the application is not in the foreground. The application can be installed on the phone without any elevated privileges and operates entirely in user space. Ballooning must maintain a minimum of 5% of the erase blocks free to avoid perpetual warnings about low free space. Purging triggers a brief warning about low free space that disappears when purging completes.

We now present the experiments we performed using ballooning on simulated flash media of different sizes. We varied the amount of ballooning that was performed and measured the time that discarded data remained on the storage medium to determine ballooning's effectiveness. We measured the ratio of deleted pages on erase blocks, which intuitively captures the amount of ballooning. We also measured the rate of flash erase block allocations, which intuitively captures the added cost of ballooning. After each simulation execution, we performed purging and measured the additional erase block allocations, which is the purging cost for the amount of ballooning used by our hybrid solution.

The erase block allocation rate tells us directly the rate that pages are written to the flash storage medium. Data can be written from two sources: the actual data written by the simulator, and the data copied by the log-structured file system's compactor. Our simulator uses a constant write distribution and therefore the expected rate of writes from the simulator is the same for all experiments. Therefore, the observed disparity in erase block allocation rates reflects exactly the additional writes resulting from the increased compactions caused by our application to achieve secure deletion.

To quantify how promptly secure data deletion occurs, we measure the expected time data remains on the storage medium. We calculate this measurement using our pattern writer that periodically writes one page pattern onto the medium and deletes them. We then compute how long the written pattern remains on the storage medium.

5.6.1 Experimental Results

Table 5.2 presents the results of simulated storage media usage with different ballooning thresholds. The partition size is the full storage capacity of the medium. The fill ratio is the average proportion of valid data on erase blocks in the storage medium, ignoring both completely full and completely empty erase blocks. We compute this by taking the periodic average of all fill ratios for eligible erase blocks and averaging these measurements (weighted by time between observations) over the course of our experiment. The erase block allocations per hour is the rate that erase blocks are allocated on the storage medium, indicating the frequency of writes to the storage medium. We used the erase block allocation rate, along with an expected erase block lifetime of 10^4 erasures before becoming a bad block [66], to compute an expected storage medium lifetime in years assuming even wear levelling. The purge cost is the number of erase blocks that must be allocated to execute purging with this configuration. Two deletion latencies are provided: the median and 95th percentile, which give a good indication of the distribution. The maximum value is undefined, as ballooning provides no guarantee of secure deletion. Each experiment was run four times and we provide 95% confidence intervals for relevant measurements.

Table 5.2 Erase block (EB) allocations, storage medium lifetimes, and deletion times for the YAFFS file system.

partition type	free EBs	fill ratio	erase block allocs / hour	life (years)	purge cost (EB)	Deletion latency (hours) median	95th %ile
	603.8	20%	32.7 ± 2.3	54	1556.8	41.5 ± 2.6	46.2 ± 0.5
200 MiB	91.8	63%	53.4 ± 4.7	33	705.2	10.8 ± 1.7	14.6 ± 1.3
YAFFS	21.0	80%	95.0 ± 24.2	18	429.8	4.2 ± 0.6	6.6 ± 0.2
	15.1	84%	166.5 ± 42.5	10	357.8	2.6 ± 0.7	5.4 ± 1.5
	4487.2	7%	26.0 ± 1.0	68	7827.0	163.1 ± 7.1	169.6 ± 7.8
	254.1	40%	35.8 ± 3.4	50	1106.5	28.4 ± 4.1	33.6 ± 2.6
1 GiB	88.2	64%	59.8 ± 8.4	29	765.0	10.4 ± 0.5	16.1 ± 2.0
YAFFS	56.2	72%	70.4 ± 0.8	25	692.3	8.2 ± 0.6	12.6 ± 2.6
	26.1	82%	163.6 ± 18.9	10	525.2	4.3 ± 0.4	7.6 ± 0.6
	23.7	83%	232.9 ± 11.4	7	360.8	3.0 ± 0.4	6.1 ± 0.6
	9503.7	4%	25.3 ± 0.8	70	15663.8	349.4 ± 11.2	370.3 ± 5.9
	387.8	43%	36.6 ± 1.5	49	1630.5	34.7 ± 7.5	43.1 ± 8.6
2 GiB	254.5	48%	41.1 ± 3.7	43	1237.5	28.7 ± 1.5	34.8 ± 6.1
YAFFS	56.4	76%	87.5 ± 5.8	20	845.8	8.5 ± 0.9	13.0 ± 0.4
	37.2	80%	205.4 ± 24.3	8	484.8	4.7 ± 0.5	9.4 ± 1.9
	36.9	80%	248.2 ± 33.0	7	338.4	3.3 ± 0.7	7.4 ± 1.0

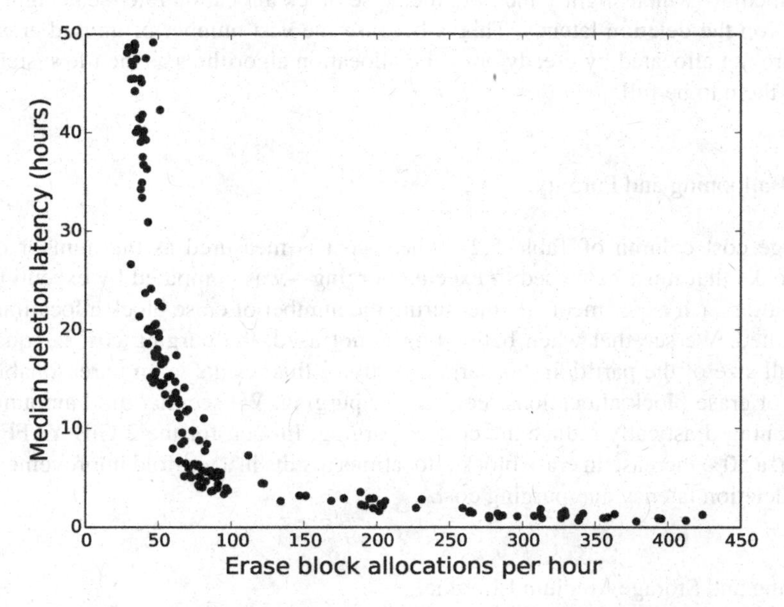

Fig. 5.5 Scatter plot of deletion latency and erase block allocation for experiments on a 200 MiB storage medium with varied ballooning.

Deletion Latency Versus Block Allocation Rate.

As discussed in Section 5.4.2, without ballooning both the fill ratios and the deletion latency are highly dependent on the size of the storage medium. As ballooning increases the fill ratio, however, the deletion latency similarly decreases. Since the data being stored comes from the same distribution, more-full erase blocks on identically sized storage media imply that there are fewer erase blocks available to store data, so the expected erase block reallocation period decreases and deleted data is removed from the system more frequently.

We observe an inverse relationship between the fill ratio and the erase block allocation rate for each partition type. Fewer available erase blocks mean more compaction and thus more frequent writes to the storage medium simply to copy data stored elsewhere. Figure 5.5 plots the relationship between the median deletion latency and the erase block allocation rate for simulations involving varying amounts of ballooning. The horizontal axis is the erase block allocation rate and the vertical axis is the median deletion time. A point on the plot represents an experiment with some amount of ballooning that resulted in the observed allocation rate and deletion latency.

The device's size is not an overriding factor in deletion latency—deletion latency can be reduced for any storage medium simply by applying the appropriate amount of ballooning to consume the excess capacity. Small amounts of ballooning on large storage media—which slightly increase the erase block allocation rate—can significantly drop the deletion latency. This is because the vast number of unused erase blocks are not allocated by greedy or cyclic allocation algorithms as the file system believes them to be full.

Hybrid Ballooning and Purging.

The purge cost column of Table 5.2—where cost is measured as the number of erase blocks that must be erased to execute purging—was computed by executing purging after each experiment and measuring the number of erase block allocations that resulted. We see that when ballooning is not used, the purging cost is equal to the full size of the partition. For large partitions, this results in an unreasonable number of erase block allocations required for purging. We see that mild amounts of ballooning drastically reduce the cost of purging. In fact, for the 2 GiB YAFFS partition, a 50% increase in erase block allocations results in a ten-fold improvement in both deletion latency and purging cost.

Ballooning and Storage Medium Lifetime.

The primary drawback of our solutions is the cost of increased erasures, both in terms of damage to the flash memory and power consumption. The additional wear is directly proportional to the increase in the erase block allocation rate, and inversely proportional to the lifespan. We compute an expected lifetime in years from the erase block allocation rate and present this in Table 5.2. We use a conservative (i.e., pessimistic) estimate of 10^4 erasures per erase block. Recall that a typical flash erase block can handle between 10^4 and 10^5 erasures [81], and some studies have indicated this is already orders of magnitude more conservative than reality [82].

Our results show that even at high erase block allocation rates, we still expect to see the storage medium live for upwards of a decade; this is well in excess of the replacement period of mobile phones that varies from two to eight years [83]. Users who require decades of longevity from their mobile phone can simply use mild ballooning. In particular, large-capacity storage media combined with mild ballooning yield a system with reasonable purging performance and flash memory lifetime reduction.

Power Consumption.

To test if our solutions have acceptable power requirements, we analyzed the power consumption of write operations. We measured the battery level of our Nexus One through the Android API, which gives its current charge as a percentage of its battery capacity. The experiment consisted of continuously writing data to the phone's flash memory in a background service while monitoring the battery level in the foreground. We measured how much data must be written to consume 10% of the total battery capacity. We ran the experiment four times and averaged the result. The resulting mean is within the range of 11.01 ± 0.22 GiB with a confidence of 95%, corresponding to 90483 full erase blocks' worth of data. Since this well exceeds the total of 1570 erase blocks on the device's data partition, we are certain that our experiment must have erased the erase blocks as well as written to them, thus measuring the power consumption of the electrically intensive erasure operation.

Even using the most aggressive ballooning measurement for YAFFS, where nearly 250 erase blocks are allocated an hour, it would take 15 days for the ballooning application's writing behaviour to consume 10% of the battery. Furthermore, the built-in battery use information reported that the testing application was responsible for 3% of battery usage, while the Android system accounted for 10% and the display for 87%. We conclude that ballooning's power consumption is not a concern.

The power consumption required for purging is related to the size of the storage medium and the capacity of the battery—0.9% of the battery per gigabyte for the Nexus One. Other mobile phone batteries may of course yield varying results. Any mobile phone with a storage medium size exceeding a gigabyte therefore consumes significant time and energy to perform purging. Our hybrid solution, however, is perfectly suited for such storage media as it significantly drops the cost of purging.

5.7 Summary

In this chapter we considered deletion latency for log-structured file systems and showed that there is no guarantee of deletion on such file systems. We presented three useful user-level solutions for secure deletion on YAFFS file systems: purging, ballooning, and a hybrid of both. The hybrid provides secure data deletion against a computationally unbounded unpredictable multiple-access coercive adversary, turning the storage medium into a SECDEL-CLOCK implementation. We have evaluated the solutions' effectiveness in terms of wear on the flash memory, as well as power consumption and time.

We restate that these solutions make strong assumptions on the implementation that stores the data, in particular that filling the capacity of the file system effects the secure deletion of all discarded data. Verifying this is simple for interfaces like MTD which provide raw access to the flash memory; however, it is not as straightforward when the memory is hidden behind an obfuscating controller.

We have also seen that user-level solutions are limited. The space of possible solutions is constrained to creating and deleting files. We showed that by filling certain log-structured file systems to capacity, we can securely delete data. It requires that the file system reclaims all wasted storage resources before proclaiming the device is full.

In the next chapter, we consider what can be achieved without a user-level access restriction and develop an efficient and prompt secure deletion solution that can be integrated into any flash file system.

5.8 Research Questions

- Are there FTLs implementations for which purging always works?
- Does balloon file rotation actually promote wear levelling? How much can this be influenced from user space?
- Does ballooning interfere with other flash memory optimizations, which may be FTL or file-system specific?
- Can anything be done to help purging *play fair* with other applications attempting to write once the device is full?

Chapter 6
Data Node Encrypted File System

6.1 Introduction

This chapter presents the Data Node Encrypted File System (DNEFS), a file system modification that augments a file system with efficient secure deletion. DNEFS is tailored to flash memory, which inhibits secure deletion by forcing all deletions to occur at a large granularity. Consequently, both flash file systems and flash translation layers implement a log-structured file system: one where new data is written at the log's end and logical addressing is used to locate file data. In log-structured file systems, data deletion happens only during compaction when the wasted storage capacity is needed for new data.

DNEFS encrypts each individual indivisible data node, i.e., the smallest unit of read/write for the file system.[1] Each data node is assigned a unique key and keys are colocated in a small set of erase blocks. This set is called the key storage area (KSA). While the main storage remains unchanged, the KSA must be capable of securely deleting data. The efficiency of the system comes from the size disparity between the KSA and the main storage, as only the former needs to perform the expensive deletion operation.

6.2 System and Adversarial Model

DNEFS is designed to provide secure data deletion for flash memory. As with the solutions presented in the previous chapter, we assume that the user is continually storing and retrieving sensitive data on a flash-based storage medium such as a mobile phone.

[1] Note that we use the term data node for consistent terminology with our UBIFS-based implementation in the next chapter. A data node is the same as a data item: an indivisible piece of data, which forms the minimum unit of I/O for a system.

J. Reardon, *Secure Data Deletion*, Information Security and Cryptography,
DOI 10.1007/978-3-319-28778-2_6

The adversary can perform a coercive attack at multiple points in time to gain access to the user's device. The adversary is computationally bounded and therefore cannot decrypt encrypted data without the corresponding key.

The user's goal is to provide secure deletion for as much data as possible. In particular, it is to limit the information learned by the adversary through compromise to only the valid data currently stored on the device. DNEFS approaches this goal by providing secure data deletion with configurable existential and deletion latencies.

6.3 DNEFS's Design

In this section we describe DNEFS: a file system modification that provides efficient fine-grained secure deletion for flash memory. DNEFS uses encryption to provide secure deletion. It encrypts each individual data node (i.e., data item, or the unit of read/write for the file system) with a different key, and then manages the storage, use, and secure deletion of these keys in an efficient and transparent way for both users and applications. Data nodes are encrypted before being written to the storage medium and decrypted after being read; this is all done in-memory. The keys are stored in the KSA: a reserved area of the file system. Figure 6.1 illustrates applying DNEFS to a file system and partitioning the storage medium.

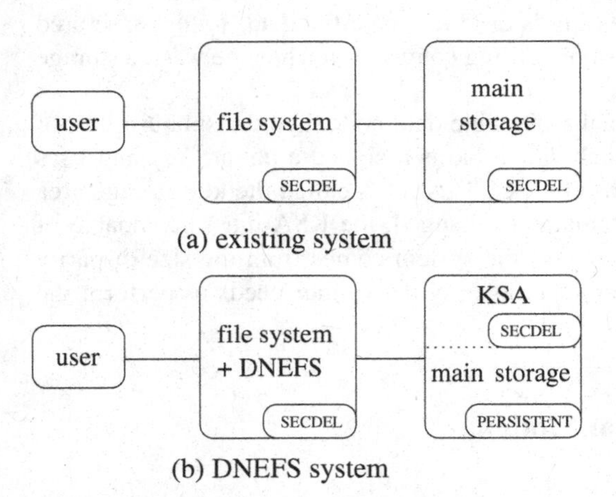

Fig. 6.1 Adding DNEFS to an existing file system. (a) The existing system consists of a SECDEL storage medium accessed by the user through a file system; deletion may be an expensive operation. (b) A DNEFS system consists of a securely deleting key storage area and a persistent main storage that constitute the storage medium. The efficiency of the system comes from applying the expensive secure-deletion operation only to the KSA.

DNEFS works independently of the notion of files; file count, file size, and file access patterns have no influence on the size of the key storage area. The encrypted data stored on the medium is no different than any reversible encoding applied by the storage medium (e.g., error-correcting codes) because all legitimate access to the data only observes the unencrypted form. This is not an encrypted file system, although in Section 6.4.5 we explain that it is easily extended to one. In our case, encryption is simply a coding technique that we apply immediately before storage to reduce the number of bits required to securely delete a data node from the data node size to the key size.

6.3.1 Key Storage Area

We assume that the storage medium is divided into erase blocks which can be atomically erased, e.g., flash memory erase blocks. Our solution uses a small set of *erase blocks* to store all the data nodes' keys—this set is called the *key storage area*. The erase blocks that are not part of the KSA belong to the main storage. The KSA is managed separately from the rest of the file system. It does not behave like a log-structured file system; instead, secure deletion is explicitly provided by using batched compactions (see Section 4.3). Our solution therefore requires that the file system or flash controller that it modifies can logically reference the KSA's erase blocks and erase old KSA erase blocks promptly after writing a new version. Figure 6.1 emphasizes that the KSA, unlike the main storage, requires secure deletion.

Each data node's header stores the logical KSA position that contains its decryption key. The erase blocks in the KSA are periodically erased to securely delete any keys that decrypt discarded data. When the file system no longer needs a data node—i.e., the data node is removed or updated—the data node's corresponding key is discarded. This data-node-based approach is independent of files; keys are discarded whenever the data node they encrypt is discarded. A key remains in the discarded state until it is securely deleted from the KSA and its location is replaced with fresh, unused, random data; its state is then changed to unused.

When a new data node is written to the storage medium, an unused key is selected from the KSA and its position is stored in the data node's header. The key is used to encrypt the data node using a symmetric key block cipher in counter mode; we use a fixed initialization vector because keys are never used to encrypt multiple data nodes. DNEFS does cryptographic operations seamlessly, so applications are unaware that their data is being encrypted. Figure 6.2 illustrates DNEFS's (a) write, (b) read, and (c) discard algorithms.

6.3.2 Keystore

The KSA is used to implement a *keystore*, which is the name we give to a special kind of storage medium whose purpose is to assign and securely delete cryptographic keys, which we call *key values* (KVs). These key values are assigned to

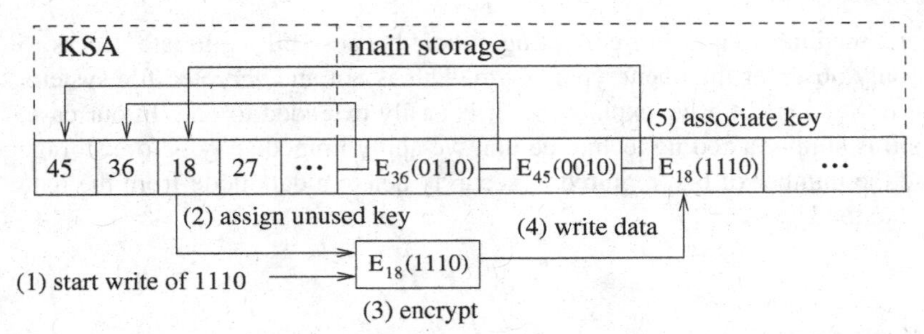

(a) DNEFS write operation

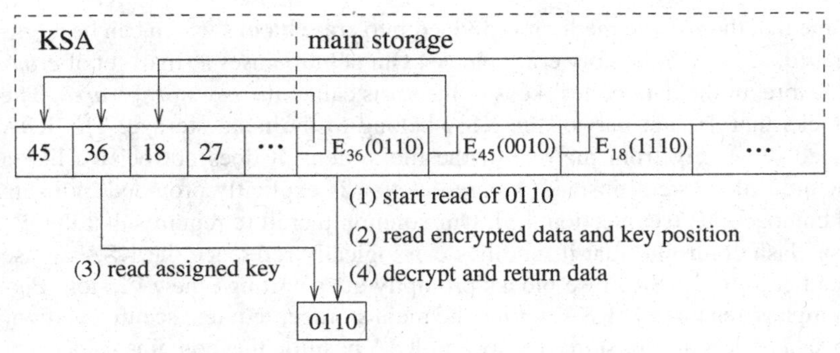

(b) DNEFS read operation

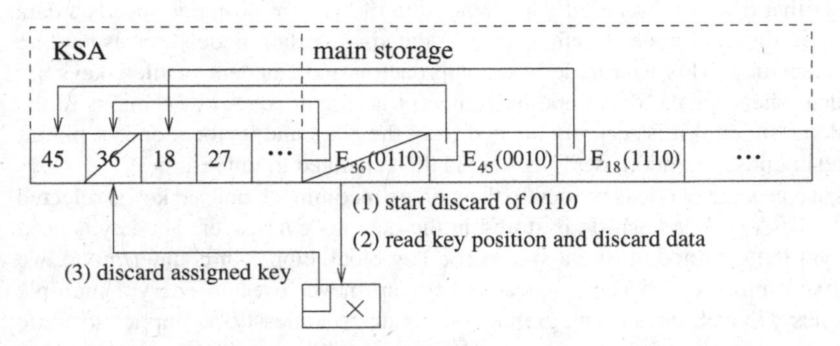

(c) DNEFS discard operation

Fig. 6.2 DNEFS's read, write, and discard datapaths. Data blocks are represented with binary strings and encryption keys are represented with two-digit numbers. (a) Writing a new data node 1110: 1110 is first encrypted with an unused key 18 and then written to an empty position in the main storage with a reference to the key's position stored alongside. (b) Reading a data node 0110: $E_{36}(0110)$ is first read from the main storage along with a reference to its key 36 in the KSA. The key is then read and used to decrypt the data. (c) Discarding a data node 0110: the key position associated to it is read, the data node and its corresponding key position are discarded.

encrypt new data and are deleted in lieu of the data they encrypt to provide secure deletion against our computationally bounded adversary.

A *keystore* has a state and three functions: `assign`, `read`, and `discard`. The keystore's state can be examined by the adversary during a coercive attack. A keystore has a security parameter κ, which is used as the length of KVs. Additionally, each KV has an access token (AT) that uniquely identifies a KV from the time its KP is assigned until it is discarded. The `assign` function takes no parameters and returns an AT or $\perp$.[2] The `read` function takes an AT and returns a KV or $\perp$. The `discard` takes an AT and returns $\top$ for success or $\perp$ for failure.

The following properties ensure that this system provides secure data deletion at a fine granularity:

- **P1** The KVs associated to the ATs returned by `assign` must be unpredictable with a negligible guessing probability decreasing exponentially in κ.
- **P2** KVs returned by `read` must be the same for a particular AT from the time `assign` returns it until it is provided to `discard`; further, `read` must not return $\perp$ during this time.
- **P3** KVs returned by `read` must be unpredictable given the keystore's state at all times before the time `assign` returns its corresponding AT minus a bounded existential latency, and at all times after the time `discard` is called with its corresponding AT plus a bounded deletion latency.

With reference to Figure 6.2, DNEFS's use of the keystore consists of three operations: (a) assigning an unused key, (b) reading an assigned key, and (c) discarding an assigned key. We now describe an example implementation of a keystore, which we call a *clocked keystore*.

6.3.3 Clocked Keystore Implementation

A clocked keystore consists of a set of key positions (KPs), each one storing a KV and a key state. The three key states are **U** for *unused*, **A** for *assigned*, and **D** for *discarded*. The assign operation provides an *unused* (**U**) KP and changes its state to *assigned* (**A**). The read operation returns the KV for an *assigned* (**A**) KP. The discard operation takes an *assigned* KP and changes its state to *discarded* (**D**).

A periodic *clock* operation securely deletes any discarded keys and returns their state to unused. During this operation, KVs for both **U** and **D** KPs are replaced with fresh, unused random data taken from a cryptographically suitable random source, e.g., `/dev/random`. Thus, unused KPs store a recently generated random KV not previously assigned; assigned KPs are those that have been assigned but not discarded; discarded KPs are those that have been assigned and then discarded, though the KV remains until it is securely deleted.

[2] In implementation the assign function returns both the AT and the KV in anticipation of an immediate read.

The keystore design achieves properties **P1–3**. **P1** holds because only **U** KPs are assignable; they store unpredictable random KVs and assigning transitions to **A**. **P2** holds because from the time a KP's state is **A** until it is **D**, the corresponding KV remains stored and is always returned. **P3** holds by replacing the KVs for both **U** and **D** KP during the clock function. The period between clock functions therefore forms an upper bound on the possible existential and deletion latencies.

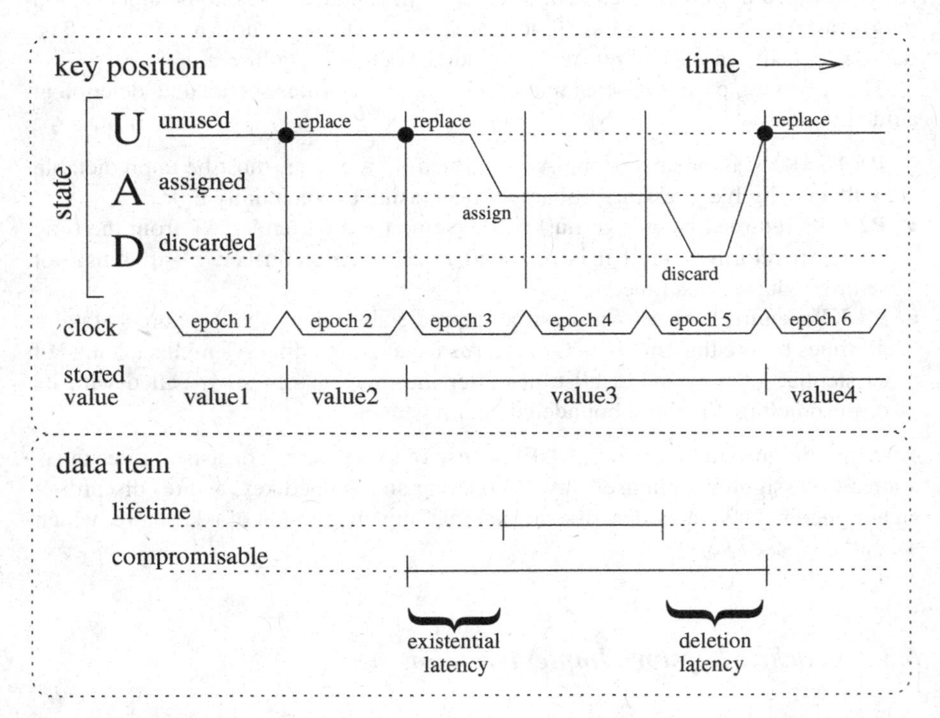

Fig. 6.3 Example timeline for a KP's state and value in a clocked keystore as well as a data node that uses the KV. The three states are unused, assigned, and discarded. A solid line plots a function over time going among these states: starting unused, then being assigned in epoch three, discarded in epoch five, and finally returning to unused. The clock triggers periodically resulting in a vertical stroke in the state. Black dots and a *replace* label indicate the creation of a new KV. Over six deletion epochs, four different KVs occupy the KP. One KV, `value3`, is contained at the time the data node is created. It is assigned and discarded at the data node's creation and deletion times, respectively. The resulting existential latency and deletion latency are indicated in the increased time that the data node is compromisable.

Figure 6.3 shows the three states available for a KP and an example history of state transitions for a KP in a clocked keystore implementation. Observe that KVs are already stored on the storage medium before they are assigned and they remain on the storage medium for a limited amount of time after they are discarded. The periodic clock operation divides time into distinct *deletion epochs*, and the KV stored in the KP changes only at the clock. An example data item has a lifetime from the third to fifth deletion epochs; its existential and deletion latencies expand in both directions to the nearest clock edge and so the time it is compromisable increases. A compromise at any timepoint, therefore, is equivalent to a compromise at any other time point in the same epoch.

6.3.4 Clock Operation: KSA Update

DNEFS clock operation, called KSA update, replaces all **U** and **D** keys with fresh values. It executes iteratively over each of the KSA's erase blocks as follows: a new version of the erase block is prepared where the **A** keys remain in the same position and all the **U** and **D** keys are replaced with fresh random data suitable for new keys. We keep assigned keys fixed because their corresponding data node has already written its logical position in the KSA for retrieval when reading. The new version of the erase block is then written to an arbitrary empty erase block on the storage medium. After completion, all erase blocks containing old versions of the logical KSA erase block are erased, thus securely deleting the unused and discarded keys along with the data nodes they encrypt.

This implementation requires the ability to securely delete an entire erase block, i.e., perform an erase block erasure. Therefore, for flash memory, DNEFS must be implemented either into the logic of a file system that provides access to the raw flash memory (e.g., UBIFS) or into the logic of the flash controller (e.g., FTL controller). Note that while both the KSA and the main storage are colocated on the same storage medium, it is only the KSA that has any secure deletion requirements; the main storage is assumed to be persistent. The efficiency of DNEFS comes from the fact that only the small number of KSA erase blocks must be erased to securely delete all data nodes that are discarded since the previous clock. This comes at the cost of assuming a computationally bounded adversary—an information-theoretic adversary could decrypt the encrypted file data.

6.3.5 Key-State Map

The KPs' states are managed in memory with a *key-state map*. Figure 6.4 shows an example key-state map and a corresponding KSA before and after a KSA update: **U** and **D** KPs are replaced with new KVs; **A** KPs retain their KVs.

When the file system is mounted, the key-state map must be correctly constructed. The actual procedure to do this depends on the file system in which it is integrated, but it must account for the possibility of a previous unsafe unmounting. We define a *correct* key-state map as one that has (with cryptographically high probability) the following three properties:

- **C1** Every unused key must not decrypt any data node—either valid or invalid.

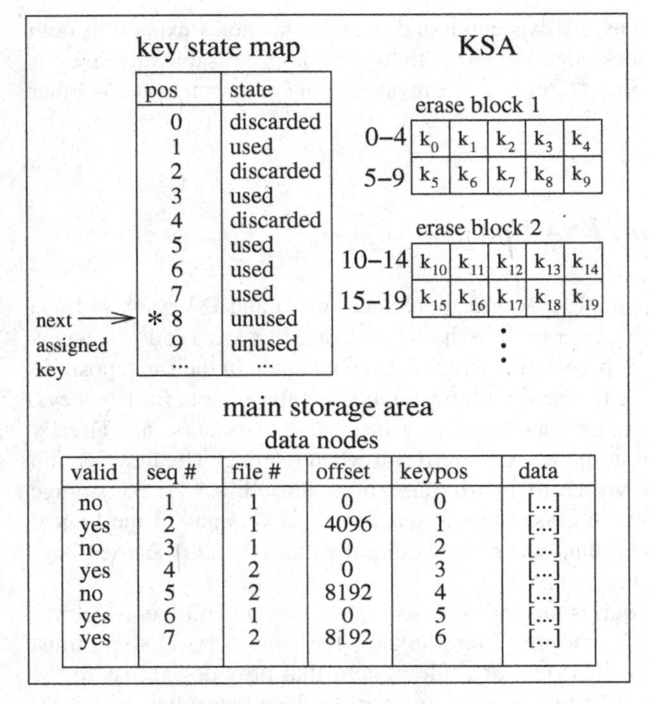

pos	state
0	discarded
1	used
2	discarded
3	used
4	discarded
5	used
6	used
7	used
* 8	unused
9	unused
...	...

valid	seq #	file #	offset	keypos	data
no	1	1	0	0	[...]
yes	2	1	4096	1	[...]
no	3	1	0	2	[...]
yes	4	2	0	3	[...]
no	5	2	8192	4	[...]
yes	6	1	0	5	[...]
yes	7	2	8192	6	[...]

(a) state before purging keys

key state map

KSA

pos	state
* 0	unused
1	used
2	unused
3	used
4	unused
5	used
6	used
7	used
8	unused
9	unused
...	...

next assigned key

erase block 1

0–4 k'_0 k_1 k'_2 k_3 k'_4

5–9 k_5 k_6 k_7 k'_8 k'_9

erase block 2

10–14 k'_{10} k'_{11} k'_{12} k'_{13} k'_{14}

15–19 k'_{15} k'_{16} k'_{17} k'_{18} k'_{19}

(b) state after purging keys

Fig. 6.4 Example of a key-state map, key-storage area, and main-storage area during a KSA update. (a) shows the state before and (b) shows the state after updating. Some keys are replaced with new values, corresponding to data nodes that were unused or discarded. The table of data nodes illustrates a log-structured file system, where newer versions of data nodes for the same file/offset invalidate older versions.

- **C2** Every assigned key must have exactly one data node it can decrypt and this data node must be referred to by the file system's index.
- **C3** Every discarded key must not decrypt any data node that is referred to by the file system's index.

Observe that an unused key that is marked as discarded still results in a correct key-state map, as it affects neither the security of discarded data nor the availability of valid data.

We note that it is always possible to build a correct key-state map. By design, file systems are capable of generating (some representation of) a file system index data structure that maps each valid data node to the location of its most-recently stored version. To build a correct key-state map, we require that for each data node in the index, its corresponding KP's state is **A**. This approach, however, requires enumerating all data nodes. In the next chapter we show that our implementation of DNEFS for UBIFS leverages UBIFS's commit and replay mechanism to greatly improve the performance of rebuilding a correct key-state map.

6.3.6 Summary

DNEFS provides guaranteed secure deletion against a computationally bounded unpredictable multiple-access coercive attacker. When an encryption key is securely deleted, the data it encrypted is then inaccessible, even to the user. All discarded data nodes have their corresponding encryption keys securely deleted during the next KSA update. KSA updates occur as a periodic clock operation, so during operation the deletion latency *for all* data is bounded by the clock period. Neither the key nor the data node is available in any deletion epoch prior to the one in which it is written, so the existential latency *for all* data is also bounded by the clock period.

6.4 Extensions and Optimizations

In this section we present some extensions to DNEFS that may improve performance or security.

6.4.1 Granularity Trade-off

DNEFS encrypts each data node with a separate key using a symmetric key block cipher in counter mode; this allows efficient secure deletion of data from long-lived files, e.g., databases. Other related work instead encrypts each file with a unique key, allowing secure deletion only at the granularity of an entire file [77]. This is well suited for media files, such as digital audio and photographs, which are usually

Table 6.1 Data node granularity trade-offs assuming 64 2-KiB pages per erase block.

data node size (KiB)	pages per data node	KSA size (EBs per GiB)	copy cost (EBs)
2	1	64	0.0
4	2	32	0.016
8	4	16	0.047
16	8	8	0.11
32	16	4	0.23
64	32	2	0.48
128	64	1	0.98

created, read, and discarded in their entirety. Using a single key per file, however, means that modifications to files require re-encrypting its entire contents with a new key and securely deleting the old key. This cost grows with the size of the file, and then becomes more efficient to use naive compaction for files larger than an erase block.

We note that random read access can be made efficient by storing periodic initialization vectors (IVs) for long files.[3] Effectively, DNEFS does this by eschewing IVs altogether and using the storage to instead store encryption keys, which, unlike IVs, can also be used for secure deletion purposes as well as efficient random access.

Thus, a trade-off exists between the storage costs of keys and the copying costs for modifications. At one extreme, DNEFS stores one key per data node and allows modifications with no additional cost. At the other extreme, one key per file (or storage medium) requires minimal storage but modifications are expensive. Between these extremes lies a range of possible encryption granularities, e.g., one key every eight data nodes.

Table 6.1 compares the encryption granularity trade-off for a flash drive with 64 2-KiB pages per erase block. To compare DNEFS with schemes that encrypt each file separately, simply consider the data node size as equal to the IV granularity or the expected file size. The KSA size, measured in erase blocks per GiB of storage space, is the amount of storage required for IVs and keys, and is the worst-case number of erase blocks that must be erased during each KSA update. The copy cost, also measured in erase blocks, is the amount of data that must be re-written to the flash storage medium due to a data node modification that affects only one page of flash memory. For example, with a data node size of 16 KiB and a page size of 2 KiB, the copy cost for a small change to the data node is 14 KiB. This is measured in erase blocks because the additional writes, once filling an entire erase block, result in an additional erase block erasure that is otherwise unnecessary with a smaller data node size. Observe that in the final row, the data node size equals the erase block size and consequently any small change requires rewriting an erase block's worth of data. In this case, however, the file system should instead store each

[3] For completeness we mention (but do *not* advise) that a cipher in electronic codebook mode permits random access at the steep cost of the loss of semantic security.

data node on its own erase block and perform erase block erasure whenever a new version is written.

6.4.2 KSA Update Policies

While DNEFS uses batching to improve efficiency, there is no technical reason that prohibits immediate secure deletion. KSA update can be automatically triggered, for example, if data from a file marked with a *sensitive* attribute is discarded. KSA update is also triggered by an `ioctl`, which means that users or applications can force its operation, e.g., after clearing the web browsing cache. Note that batching of discards is required for DNEFS to provide any benefits over naive compaction of the erase block containing discarded data.

In addition to periodic updates, KSA updates can also be triggered once the number of discarded keys exceeds a threshold; this ensures that both the deletion latency and the amount of exposable data is limited. This effects a natural user interface, where discarding many files triggers secure deletion in the same way that a full garbage bin causes it to be emptied.

6.4.3 KSA Organization

The KSA can be divided into groups with different properties. This can be to provide extra features or to improve efficiency. For example, KSA groups may vary in their clock frequency, so that sensitive data may be more quickly securely deleted. KSA groups may also vary in their encryption key sizes. We revisit this idea in Chapter 11.

Since the efficiency of DNEFS comes from batching, colocating the keys for data that is discarded simultaneously results in more efficient erase block erasures (i.e., more discarded keys per erased erase block). When the expiration time of data is not known in advance, a coarse division into short-term and long-term KSA groups can be approximated. When a data node is written to the file system it is encrypted with a short-term storage key. If the file system's free-space compaction results in that data node being moved, it can be re-encrypted with a new key from the long-term storage area. Thus, a form of generational garbage collection is used as a heuristic to identify longer-lived data and encrypt it with a long-term KSA key [84].

6.4.4 Improving Reliability

As a technical note, flash erase blocks may become unreadable or unwritable. If a KSA erase block becomes unreadable, only a few KiB of keys are lost. Unfortunately, this corresponds to the loss of a much larger amount of data. Depending on the characteristics of the flash memory, it may be appropriate to replicate the KSA to prevent the loss of any data.

If a KSA erase block becomes a bad block while erasing it, it may be possible that its contents remain readable on the storage medium without the ability to remove them [66]. In this case, it is necessary to re-encrypt any data node whose encryption key remains available and to force the compaction of those erase blocks on which the data nodes reside. More generally, the implementation of the keystore as a set of KSA erase blocks does not guarantee robustness in data storage: one that always stores data correctly and always securely deletes data correctly. In Chapter 11 we show how to make a keystore that is robust against partial failures in confidentiality, integrity, and availability.

6.4.5 Encrypted File System

Our design can be trivially extended to offer a passphrase-protected encrypted file system: we simply encrypt the KSA whenever we write random data with a key derived from a password-based key derivation function, e.g., similar to LUKS [85].

Because each randomly generated key in the KSA is unique (with high probability), we can encrypt the KSA using a block cipher in ECB mode to allow rapid decryption of randomly accessed offsets without storing additional initialization vectors [86]. Provided that the ciphertext block size is the same as the encryption key, no data is needlessly decrypted.

6.5 Summary

DNEFS is a generic file system extension designed for adding secure deletion to data and is particularly suited to flash memory. It provides secure deletion against a *computationally bounded unpredictable multiple-access coercive adversary*, turning the storage medium into a SECDEL-CLOCK-EXIST implementation.

DNEFS works by encrypting each data node with a different key and storing the keys together on the flash storage medium. The erase blocks containing the keys are periodically updated to remove old keys, replacing them with fresh random data that can be used as keys for new data. DNEFS provides *fine-grained* deletion in that parts of files that are overwritten are also securely deleted.

In the next chapter, we describe UBIFSec, which is an implementation of DNEFS for the flash file system UBIFS. We further deploy UBIFSec on an Android mobile phone and test its performance in practice to verify that it is efficient.

6.6 Research Questions

- How should the KSA be best organized to colocate keys with approximate expiration times so as to minimize erase block erasures? How do file system workloads affect KSA organization optimization?
- How should a securely deleting cache for flash memory be designed? Note that simply encrypting it is unsuitable for secure deletion if the cache persists for a long time, however the storage of keys on disk may not be necessary—provided there are not too many kept in RAM.
- DNEFS's storage of key positions in the data node may leak novel metadata. Assigning KPs sequentially, for instance, may reveal data nodes belonging to the same file or created at the same time. What information is leaked, and are further mitigations required? Is random KSA key assignment sufficient, and how does this affect KSA optimization?

Chapter 7
UBIFSec: Adding DNEFS to UBIFS

7.1 Introduction

The previous chapter presents DNEFS, a generic file system extension that provides efficient secure deletion. This chapter validates DNEFS by building and testing UBIFSec: the implementation of DNEFS for the flash file system UBIFS. We measure the increased flash memory wear caused by DNEFS as well as the battery consumption and conclude that UBIFSec has excellent performance and efficiently solves the problem of secure deletion for flash memory.

DNEFS is easily integrated into UBIFS with changes to about 100 lines of existing UBIFS source code and the inclusion of a new component, the KSA. We deploy UBIFSec on a Google Nexus One smartphone [65] running an Android OS. The system and applications (including video and audio playback) run normally on top of UBIFSec.

7.2 System and Adversarial Model

This chapter focuses on a concrete instantiation of the general solution DNEFS, whose design is described in Chapter 6. UBIFSec uses the same system and adversarial model as the one described in Section 6.2.

7.3 Background

Before detailing the integration of DNEFS with UBIFS, we first provide the necessary background information. We briefly recall the MTD layer, describe a logical interface for it called UBI, and then introduce the UBI-based flash file system

J. Reardon, *Secure Data Deletion*, Information Security and Cryptography,
DOI 10.1007/978-3-319-28778-2_7

UBIFS. Recall that Figure 2.1 shows the layers and interfaces involved in accessing flash memory.

7.3.1 MTD and UBI Layers

On Linux, flash memory is accessed through the Memory Technology Device (MTD) layer [67]. MTD has the following interface: read a page, write a page, erase an erase block, check if an erase block is bad, and mark an erase block as bad. Erase blocks are referenced sequentially, and pages are referenced by the erase block number and offset.

Unsorted Block Images (UBI) is an abstraction of MTD, where logical erase blocks are transparently mapped to physical erase blocks [29]. UBI's logical mapping implements wear-levelling and bad block detection, allowing UBI file systems to ignore these details. UBI also permits the atomic updating of a logical erase block—the new data is either entirely available or the old data remains.

UBI exposes the following interface: read and write to a logical erase block (LEB), erase an LEB, and atomically update the contents of an LEB. UBI LEBs neither become bad due to wear, nor should their erasure counts be levelled. Each UBI LEB has a unique number that orders the LEBs.

Underlying this interface is an injective partial mapping from LEBs to physical erase blocks (PEBs), where PEBs correspond to erase blocks at the MTD layer. The lower half of Figure 7.1 illustrates this relationship. Wear monitoring is handled by tracking the erasures at the PEB level, and a transparent remapping of LEBs occurs when necessary. Remapping also occurs when bad blocks are detected. Despite remapping, an LEB's number remains constant, regardless of its corresponding PEB.

Atomic updates of LEBs occur by invoking UBI's update function, passing as parameters the LEB number to update along with a buffer containing the desired contents. An unused and empty PEB is selected and the page-aligned data is then written to it. UBI then updates the LEB's mapping to the new PEB, and the previous PEB is queued for erasure. This erasure can be done either automatically in the background or immediately with a blocking system call. If the atomic update fails at any time—e.g., because of a power loss—then the mapping is unchanged and the old PEB is not erased.

7.3.2 UBIFS

The UBI file system, UBIFS [75], is designed specifically for UBI, and Figure 7.1 illustrates UBIFS's relationship to UBI and MTD. UBIFS divides file data into fixed-sized data nodes. Each data node has a header that stores the data's inode number and its file offset. This inverse index is used by UBIFS's compactor (called the

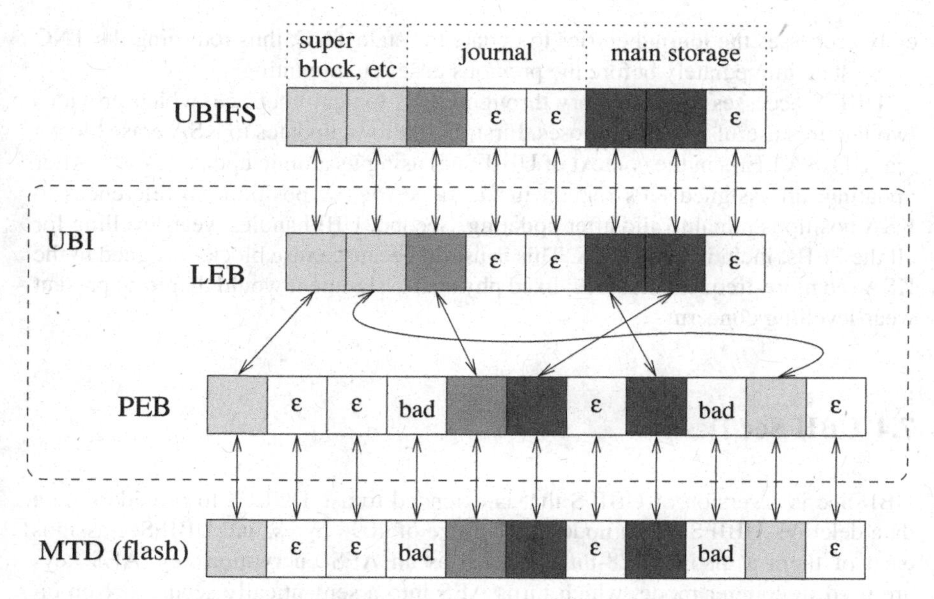

Fig. 7.1 Erase block relationships among MTD, UBI, and UBIFS. Different shades label different areas of the file system: the super block, journal, main storage, etc. Empty LEBs are labelled by ε and are not mapped to a corresponding PEB by UBI. Similarly, bad PEBs are labelled and not mapped onto by UBI.

garbage collector) to determine if the nodes on an erase block are valid or can be discarded.

UBIFS writes all new data in a journal similar to a log-structured file system; the journal consists of a set of LEBs. When the UBIFS journal is full, it is committed to the main storage area and emptied by logically moving the journal to an empty set of LEBs and growing the main storage area to encompass the old journal. An index is used to locate data nodes, and this index is also written to the storage medium. At its core, UBIFS is a log-structured file system; in-place updates are not performed. As such, UBIFS does not provide guaranteed secure data deletion.

UBIFS uses an index to determine which version of data is the most recent. This index is called the tree node cache (TNC), and it is stored both in volatile memory and on the storage medium. The TNC is a B+ search tree [87] that has a small entry for every data node in the file system. When data is appended to the journal, UBIFS updates the TNC to reference its location. UBIFS implements truncations and deletions by appending special non-data nodes to the journal. When the TNC processes these nodes, it finds the range of TNC entries that correspond to the truncated or deleted data nodes and removes them from the tree.

UBIFS uses a commit-and-replay mechanism to ensure that the file system can be mounted after an unsafe unmounting without scanning the entire device. Commit periodically writes the current TNC to the storage medium, and starts a new empty journal. Replay loads the most-recently stored TNC into memory and chronologically processes the journal entries to update the stale TNC, thus returning the TNC to the state immediately before the previous unsafe unmounting.

UBIFS accesses flash memory through UBI's logical interface, which provides two features useful for our purposes. First, UBI allows updates to KSA erase blocks (called KSA LEBs in the context of UBIFSec) using its atomic update feature. After updating, all assigned KPs remain in the same *logical* position, so references to KSA positions remain valid after updating. Second, UBI handles wear-levelling for all the PEBs, including the KSA. This is useful because erase blocks assigned to the KSA see more-frequent erasures; fixed physical assignment would therefore present wear-levelling concerns.

7.4 UBIFSec Design

UBIFSec is a version of UBIFS that is extended to use DNEFS to provide secure data deletion. UBIFS's data nodes have a size of 4096 bytes, and UBIFSec assigns each of them a distinct 128-bit KV used as an AES encryption key. AES keys are used in counter mode, which turns AES into a semantically secure stream cipher [86]. Since each AES key is only ever used to encrypt a single block of data, we can safely omit the generation and storage of initialization vectors (IVs) and simply start the counter for each AES key at a static value. Our solution requires about 0.4% of the storage medium's capacity for the KSA, although there exists a trade-off between the KSA's size and the data node's size (see Section 6.4.1).

7.4.1 Key Storage Area

The KSA is composed of a set of LEBs that store random data used as encryption keys. When the file system is created, cryptographically suitable random data is written from a hardware source of randomness to each of the KSA's LEBs, and all the KPs are marked as unused. The KSA update writes new versions of the KSA LEBs using UBI's atomic update feature; immediately afterwards, `ubi_flush` is called to ensure that all PEBs containing old versions of the LEB are synchronously erased. All KVs they contain are therefore securely deleted. This flush feature ensures that all copies of LEBs made through internal wear-levelling are also securely deleted. Figure 7.2 shows the LEBs and PEBs during a KSA update.

Only KSA erase blocks with discarded data are updated, though erase blocks that are not updated are not used to assign new KVs. To further reduce the number of KSA erase blocks that must be updated, we use KSA groups to concentrate KVs for long-term data. Our implementation uses two KSA groups: a short-term group and a long-term group. New data nodes initially get a short-term KV. If a data node is ever compacted by UBIFS, it is re-encrypted with a KV assigned from the long-

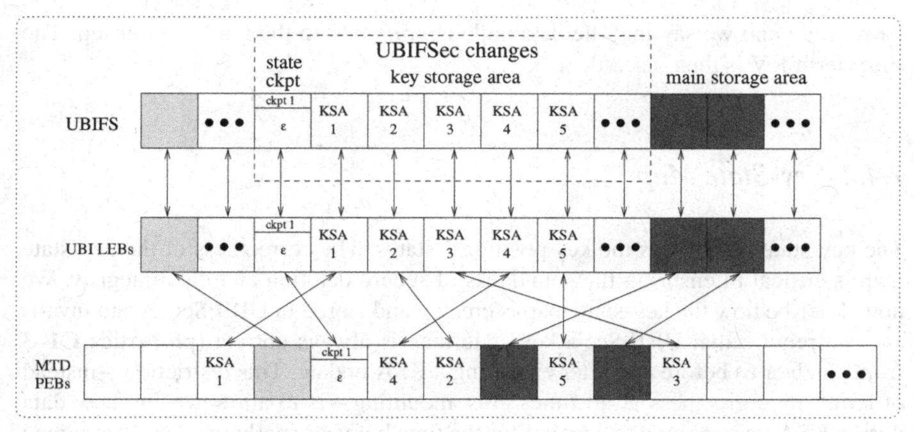

(a) before update

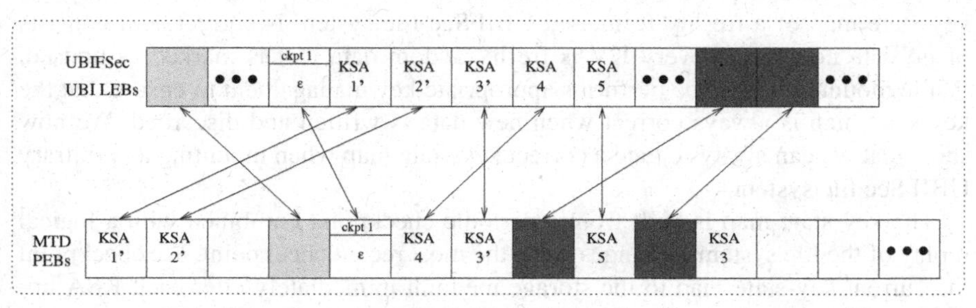

(b) during update

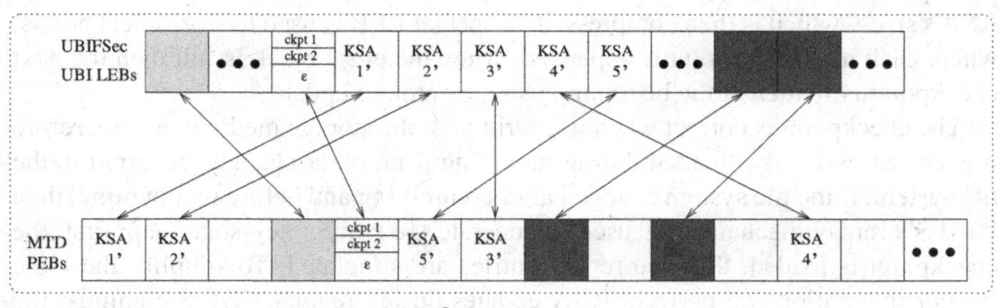

(c) after update

Fig. 7.2 Erase block relationships among MTD, UBI, and UBIFSec, showing the new regions added by UBIFSec (cf. Figure 7.1). This figure has three components that illustrate the state (a) before, (b) during, and (c) after a KSA update. Observe in (b) that new versions of KSA blocks 1, 2, and 3 are written to new locations; the old version of block 3 remains. Observe in (c) that no old KSA erase block remains and a new key-state checkpoint is written.

term group and we say that the data node is *promoted* to the long-term group. The short-term KV is then discarded.

7.4.2 Key-State Map

The key-state map stores the key positions' states. The correctness of the key-state map is critical in ensuring the soundness of secure deletion and data integrity. We now describe how the key-state map is created and stored in UBIFSec. As an invariant, we require that UBIFSec's key-state map is always correct (properties **C1–3** from Chapter 6) before and after executing a KSA update. This restriction—instead of requiring correctness at all times after mounting—is to allow writing new data during KSA updates, and to account for the time between marking a key as assigned and writing the data it encrypts onto the storage medium.

The key-state map is stored, used, and updated in volatile memory. Initially, the key-state map of a freshly formatted UBIFSec file system is correct as it consists of no data nodes, and every KV is fresh, random data that is marked as unused. While mounted, UBIFSec performs appropriate key management to ensure that the key-state map is always correct when new data is written and discarded. We now show that we can always create a correct key-state map when mounting an arbitrary UBIFSec file system.

The key-state map is built from a periodic checkpoint combined with a logical replay of the file system's changes since the most recent checkpoint. We checkpoint the current key-state map to the storage medium immediately after each KSA update. (This is even before logically replaying cached changes that occurred while updating.) After the KSA update, every key is either unused or assigned, and so a checkpoint of this map can be stored using one bit per key—less than 1% of the KSA's size—which is then compressed. A special LEB is used to store checkpoints, where each new checkpoint is appended; when the erase block is full then the next checkpoint is written at the beginning using an atomic update.

The checkpoint is correct when it is written to the storage medium, and therefore it is correct when it is loaded during mounting if no other changes occurred in the file system. If the file system changed after committing and before unmounting, then UBIFS's replay mechanism is used to generate the correct key-state map: first, the checkpoint is loaded, then the replay entries are simulated. To simplify the logic for our integration, we perform KSA updates during regular UBIFS commits; the nodes that are then replayed for UBIFS are exactly the ones that must be replayed for DNEFS. If the stored checkpoint gets corrupted, then a full scan of the valid data nodes rebuilds the correct key-state map. A consistency check for the file system also confirms the correctness of the key-state map with a full scan.

As it is possible for the storage medium to fail during the commit operation (e.g., due to a loss of power), we now show that our invariant holds regardless of the condition of unmounting. Figure 7.3 shows a flow chart of the UBIFSec commit operation, annotated with the locations where it may fail. Each action in a rectangle is

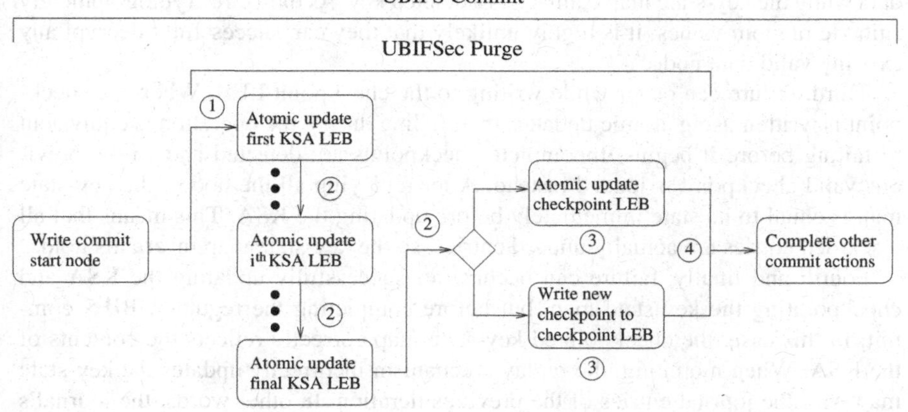

Fig. 7.3 Flow chart of UBIFSec commit process labelled with four distinct potential failure locations. Each rounded rectangle contains an action in the process that either succeeds or fails atomically. Numbers indicate the unique failure points in our analysis.

Table 7.1 Consequences of replaying false information during committing.

prev stored ckpt	possible journal	current stored ckpt	state after double replay	correct key state map
U	∅	U	U	yes
U	U→A	A	A	yes
U	U→A→D	U	D	yes
A	∅	A	A	yes
A	A→D	U	D	yes

atomic: it either succeeds or fails entirely. KSA update consists of atomically updating each LEB containing discarded KPs and afterwards writing a new checkpoint. UBI's atomic update feature ensures that any failure before completing the update is equivalent to failing immediately before beginning. Therefore, with reference to the numerical labels in Figure 7.3, the following is the complete list of distinct failure points: (1) before the first LEB update, (2) between some LEB updates, (3) after all the LEB updates but before or during the key state map checkpoint, (4) after the checkpoint but before finishing other UBIFS commit actions. We now discuss each of these failure points in detail.

First, failure can occur before updating the first LEB, which means the KSA is unchanged. When remounting the device, the loaded checkpoint is updated with the replay data, thereby constructing the exact key-state map before updating—correct by assumption.

Second, failure can occur after updating one, several, or all of the KSA's LEBs. When remounting the device, the loaded checkpoint merged with the replay data reflects the state before the first update, so some updated LEBs contain new unused data while the key-state map claims it is a deleted key. As these are cryptographically suitable random values, it is highly unlikely that they can successfully decrypt any existing valid data node.

Third, failure can occur while writing to the checkpoint LEB. When the checkpoint is written using atomic updates, then failing during the operation is equivalent to failing before it begins. Incomplete checkpoints are detected and so the previous valid checkpoint is loaded instead. After replaying all the nodes, the key-state map is equal to its state immediately before updating the KSA. This means that all discarded entries are actually unused entries, so the key-state map invariants hold.

Fourth and finally, failure can occur after successfully updating the KSA and checkpointing the key-state map, but before completing the regular UBIFS commit. In this case, the checkpointed key-state map correctly reflects the contents of the KSA. When mounting, the replay mechanism incorrectly updates the key-state map with the journal entries of the previous iteration. In other words, the journal's contents are doubly applied to the key-state map. Table 7.1 shows the space of possibilities when replaying old changes on the post-updated checkpoint; it omits impossible checkpoint-journal combinations. For all possible double-replay scenarios, the generated key-state map is always correct.

In summary, the correctness of the key-state map before and after KSA updates is invariant, regardless of when or how the file system was unmounted. This ensures secure deletion's soundness as well as valid data's integrity on the storage medium.

7.4.3 Summary

UBIFSec instantiates DNEFS for UBIFS, and so it provides efficient fine-grained guaranteed secure deletion. UBIFSec is efficient in storage space: the overhead for keys is fixed and it needs less than 1% of the total storage medium's capacity. The periodic checkpointing of UBIFSec's key-state map ensures that UBIFS's mounting time is not significantly affected by our approach.

Our implementation of UBIFSec is available as a Linux kernel patch for version 3.2.1 [88]. Table 7.2 lists the small amount of changes to the original UBIFS source code required to integrate our solution. The keystore's implementation constitutes most of the implementation effort.

7.5 Experimental Validation

We patched a Nexus One smartphone's Linux kernel to include UBIFSec and modified the phone to use it as the primary data partition. In this section, we describe experiments with our implementation on both the Android mobile phone and on a simulator.

Table 7.2 Changes to UBIFS source code required to integrate UBIFSec.

Mounting (25 lines of code)
mount the file system
- allocate and initialize the keystore
- deallocate keystore if an error occurs
- read the size of the KSA from the master node
unmount the file system
- deallocate the keystore
create default file system
- use storage medium's geometry to compute the required KSA size
- store this information in the master node
- call keystore's initialize KSA routine
Commit (3 lines of code)
commit the journal
- call the keystore's update operation
Input/Output (21 lines of code)
write data
- obtain an unused key position from the keystore
- store the key's position in the data node's header
- use the keystore and key position to look up the key
- provide the key to the compress function
recompute data after truncation
- obtain the original key, decrypt the data
- obtain a new key, encrypt the data with it after truncating
read data
- use the keystore and data node's key position to look up the key
- provide the key to the decompress function
Tree Node Cache (42 lines of code)
add/update the TNC
- provide a key position when adding data nodes
- store the key position inside TNC entries
- assign key position
- if updating, discard old key position as discarded
delete/truncate the TNC
- when removing a data node from the TNC, discard key position
commit the TNC
- read and write key position to stored tree nodes
Garbage Collection (13 lines of code)
promote key
- decide whether to promote data node
- re-encrypt promoted data node
- discard old key, assign new key

Our experiments measure our solution's cost: additional battery consumption, wear on the flash memory, and time required to perform file operations. The increase in flash memory wear is measured using a simulator, and the increase in time is measured on the Google Nexus One smartphone by instrumenting the source code of UBIFS and UBIFSec to measure the time it takes to perform basic file system operations. We further collected timing measurements from the same smartphone running YAFFS: the flash file system used on Android phones at the time that we undertook this research.

7.5.1 Android Implementation

To test the feasibility of our solution on mobile devices, we port UBIFSec to the Android OS. The Android OS is based on the Linux kernel, and it is straightforward to add support for UBIFS. The UBIFS source code is already available so we apply our patch (backporting it for Linux kernel version 2.6.35.7) and configure the kernel compiler to include the UBI device and the UBIFS in compilation. We modify the Android boot image to create UBI devices from the data partition's MTD device and mount the data partition as file system type UBIFS. Because the default file system for this Android version is YAFFS, some of our experiments compare UBIFS not only to UBIFSec but also to YAFFS.

7.5.2 Wear Analysis

We measure UBIFSec's wear on the flash memory in two ways: the number of erase cycles that occurs on the storage medium, and the distribution of erasures over the erase blocks. To reduce the wear, it is desirable to minimize the number of erasures that are performed, and to evenly spread the erasures over the storage medium's erase blocks.

We instrument both UBIFS and UBIFSec to measure PEB erasure frequency during use. We vary UBIFSec's KSA update period and compute the resulting erase block allocation rate. We do this with a low-level control (`ioctl`) that forces UBIFS to perform a commit. We also measure the expected number of deleted keys and updated KSA LEBs during KSA updates.

Using `nandsim` we simulate in memory a UBI storage medium using the geometry of Nexus One's flash memory [65]. We vary the period between UBIFSec's updates, i.e., the duration of a deletion epoch: one of 1, 5, 15, 30, and 60 minutes. We use the discrete event simulator based on the observed writing behaviour from the data collected earlier (see Section 5.4.1). Writing is performed until the file system begins compaction; henceforth we take measurements for a week of simulated time. We averaged the results from four attempts and computed 95% confidence intervals.

Table 7.3 Wear analysis for our modified UBIFS. The expected lifetime is based on the Google Nexus One's flash specifications, which have 1571 erase blocks with a (conservative) lifetime estimate of 10^4 erasures.

update period	**erased PEBs per hour**	**erasures per KSA update**	**KSA LEBs updated per hour**	**discarded KVs per updated LEB**	**wear ineq (%)**	**life (years)**
Std. UBIFS	21.3 ± 3.0	-	-	-	16.6 ± 0.5	841
60 minutes	26.4 ± 1.5	6.8 ± 0.5	6.8 ± 0.5	64.2 ± 9.6	17.9 ± 0.2	679
30 minutes	34.9 ± 3.8	5.1 ± 0.6	9.7 ± 2.0	50.3 ± 9.5	17.8 ± 0.3	512
15 minutes	40.1 ± 3.6	3.7 ± 0.4	14.9 ± 1.6	36.3 ± 8.2	19.0 ± 0.3	447
5 minutes	68.5 ± 4.4	2.6 ± 0.1	30.8 ± 0.7	22.1 ± 4.3	19.2 ± 0.5	262
1 minute	158.6 ± 11.5	1.0 ± 0.1	61.4 ± 4.6	14.1 ± 4.4	20.0 ± 0.2	113

To determine whether our solution negatively impacts UBI's wear levelling, we perform the following experiment. Each time UBI unmaps an LEB from a PEB (thus resulting in an erasure) or atomically updates an LEB (also resulting in an erasure), we log the erased PEB's number. From this data, we compute the PEBs' erasure distribution.

To quantify wear-levelling, we use the Hoover economic wealth inequality indicator [89]—a metric that is independent of the storage medium size and erasure frequency. This metric comes from economics, where it quantifies the unfairness of wealth distributions. It is equal to the normalized sum of the difference of each measurement to the mean. For our purposes, it is the fraction of erasures that must be reassigned to other erase blocks to obtain completely even wear. Let the observations be $c_1, \ldots, c_n$, and $C = \sum_{i=1}^{n} c_i$, then the inequality measure is $\frac{1}{2}\sum_{i=1}^{n} |\frac{c_i}{C} - \frac{1}{n}|$.

Table 7.3 presents the results of our experiment. We see that the rate of block allocations increases as the KSA update period decreases, with 15 minutes providing a palatable trade-off between the additional wear and timeliness of deletion. The KSA's update rate is computed as the product of the KSA update frequency and the average number of KSA LEBs that are updated each time. As such, it does not include the additional costs of executing UBIFS commits, which are captured by the disparity in the block allocations per hour. We see that when committing each minute, the additional overhead of committing compared to the updates of KSA blocks becomes significant.

As a remedy, we argue that while we integrated KSA update with UBIFS's commit to simplify the recovery logic, it is possible to separate these operations. Indeed, UBIFSec can add `KSA-update start` and `KSA-update finish` nodes as regular non-data journal entries. The replay mechanism is then extended to correctly update the key state map while processing these update nodes.

The expected number of KVs deleted per updated KSA LEB decreases sublinearly with the update period and linearly with the number of updated LEBs. This is because a smaller interval results in fewer expected deletions per interval and fewer deleted keys.

Finally, UBIFSec affects wear-levelling slightly. The unfairness increases with the update frequency, likely because the set of unallocated PEBs is smaller than the set of allocated PEBs; frequent updates cause unallocated PEBs to suffer more erasures. However, the effect is slight. It is certainly the case that the additional block erasures are, for the most part, evenly spread over the device.

7.5.3 Power Consumption

To measure battery consumption over time, we disable the operating system's suspension ability, thus allowing computations to occur continuously and indefinitely. This has the unfortunate consequence of maintaining power to the screen of the mobile phone. We first determine the power consumption of the device while remaining idle over the course of two hours starting with an 80% charged battery with a total capacity of 1366 mAh. The result was nearly constant at 121 mA. We subtract this value from all other power consumption measures.

To measure read throughput and battery use, we repeatedly read a large (85 MiB) file; we mount the drive as read-only and remount it after each read to ensure that all read caches are cleared. We read the file using `dd`, directing the output to `/dev/null` and record the observed throughput. We begin each experiment with an 80% charged battery and run it for ten minutes observing constant behaviour. We choose 80% charge to simulate steady state conditions—avoiding extremal charge states.

Table 7.4 presents the results for this experiment. For all file systems, the additional battery consumption was constant: 39 mA, about one-third of the idle cost. Depending on the file system, however, that amount of power achieved a varying throughput. We therefore include in our results a computation of the amount of data that can be read using 13.7 mAh—1% of the Nexus One's battery. The write throughput and battery consumption was measured by using `dd` to copy data from `/dev/zero` to a file on the flash file system. Compression is disabled for UBIFS for comparison with YAFFS. When the device is full, the throughput is recorded. We immediately start `dd` to write to the same file, which begins by overwriting it and thus measuring the battery consumption and reduction in throughput imposed by erase block erasures concomitant with writes.

7.5.4 Throughput Analysis

Table 7.4 shows read and write throughput achieved for different file systems. We observe that the use of UBIFSec reduces the throughput for both read and write operations when compared to UBIFS. Some decrease is expected, as the encryption keys must be read from flash while reading and writing. To determine whether there is any added latency due to the cryptographic operations, we performed these exper-

Table 7.4 I/O throughput and battery consumption for YAFFS, UBIFS, and UBIFSec.

	YAFFS	UBIFS	UBIFSec
Read rate (MiB/s)	4.4	3.9	3.0
Power usage (mA)	39	39	39
GiB read per 1% battery	5.4	4.8	3.7
Write rate (MiB/s)	2.4	2.1	1.7
Power usage (mA)	30	46	41
GiB written per 1% battery	3.8	2.2	2.0

iments with a modified UBIFSec that immediately returned zeroed memory when asked to read a key, but otherwise performed all cryptographic operations correctly. The resulting throughput for read and write was identical to UBIFS, suggesting that (for multiple reads) cryptographic operations are easily pipelined into the relatively slower flash memory read/write operations.

Some key caching optimizations can be added to UBIFSec to improve the throughput. Whenever a page of flash memory is read, the entire page can be cached at no additional read cost, allowing efficient sequential access to keys, e.g., for a large file. Long-term use of the file system may reduce its efficiency as fragmented gaps between unused and assigned keys result in sequential blocks of data not being assigned sequential keys in the KSA, causing frequent cache misses for sequential reads. Improved KSA organization can help retain this efficiency.

Write throughput, alternatively, is easily improved with caching. The sequence of keys for data written in the next deletion epoch is known at update time when all these keys are randomly generated and written to the KSA. By using a heuristic on the expected number of keys assigned during a deletion epoch, the keys for new data can be kept in memory as well as written to the KSA. Whenever a key is needed, it is taken and removed from this cache while there are still keys available.

Caching keys in memory exposes UBIFSec to attacks. We ensure that all memory buffers containing keys are overwritten when the key is no longer needed during normal cryptographic operations. Caches contain keys for a longer time but are cleared during KSA update to ensure deleted keys never outlive their deletion epoch. Sensitive data stored in volatile memory by applications may remain after the data's deletion; secure memory deallocation should be provided by the operating system to ensure its irrecoverability [90].

7.5.5 Timing Analysis

We time the following file system functions: mounting/unmounting the file system and writing/reading a page. Additionally, we time the following functions specific to UBIFSec: allocation of the cryptographic context, reading the encryption key, performing an encryption/decryption, and updating a KSA LEB. We collect dozens of

Table 7.5 Timing results for various file system functions on an Android mobile phone.

file system operation	80th percentile execution time (ms)		
	YAFFS	**UBIFS**	**UBIFSec**
mount	43	179	236
unmount	44	0.55	0.67
read data node	0.92	2.8	4.0
write data node	1.1	1.3	2.5
prepare cipher	-	-	0.05
read key	-	-	0.38
encrypt	-	-	0.91
decrypt	-	-	0.94
update one LEB	-	-	21.2

measurements for updating, mounting and unmounting, and hundreds of measurements for the other operations (i.e., reading and writing). We control for the delay caused by our instrumentation by repeating the experiments instead of executing nested measurements, i.e., we time encryption and writing to a block in separate experiments.

We mount a partition of the Android's flash memory first as a standard UBIFS file system and then as UBIFSec file system. We execute a sequence of file I/O operations on the file system. We collect the resulting times and present the 80th percentile measurements in Table 7.5. Because of UBIFS's implementation details, the timing results for reading data nodes contain also the time required to read relevant TNC pages (if they are not currently cached) from the storage medium, which is reflected in the increased delay. Because the data node size for YAFFS is half that of UBIFS, we also doubled the read/write measurements for YAFFS for a fair comparison. Finally, the mounting time for YAFFS is for mounting after a safe unmount—for an unsafe unmount (e.g., a crash), YAFFS requires a full device scan, which takes several orders of magnitude longer. This difference is because YAFFS checkpoints the file system's data structures when safely unmounting and simply reads them when mounting, continuing from whence it was.

The results show an increase in the time required for each of the operations. Mounting and unmounting the storage medium continue to take a fraction of a second. Reading and writing to a data node increases by a little more than a millisecond, an expected result that reflects the time it takes to read the encryption key from the storage medium and encrypt the data. We also test for noticeable delay by watching a movie in real time from a UBIFSec-formatted Android phone running the Android OS: the video was 512x288 Windows Media Video 9 DMO; the audio was 96.0 kilobit DivX audio v2. Both the video and audio play as expected on the phone; no observable latency, jitter, or stutter is observed during playback while background processes ran normally.

Each atomic update of an erase block takes about 22 milliseconds. This means that if every KSA LEB is updated, the entire data partition of the Nexus One phone can be securely deleted in 200 milliseconds. The cost to securely delete data grows with its storage medium's size. The erasure cost for KSA updates can be reduced in a variety of ways: increasing the data node size to use fewer keys, increasing the KSA's update period, or improving the KSA's organization and key assignment strategy to minimize the number of KSA LEBs that contain deleted keys. The last technique can work alongside lazy updates of KSA LEBs that contain no deleted keys (i.e., only unused and assigned keys) so that they are only updated before being used to assign new keys to ensure freshness.

7.6 Conclusions

UBIFSec implements DNEFS for UBIFS and shows that DNEFS is a feasible solution for efficient secure deletion for flash memory operating within flash memory's specifications. It provides guaranteed periodic per-data-block secure deletion against a computationally bounded unpredictable multiple-access coercive adversary. It turns the storage medium into a SECDEL-CLOCK-EXIST implementation where the clock frequency is a trade-off between the deletion/existential latencies and the device wear.

Our experiments validate that UBIFSec has little cost. It requires a small evenly levelled increase in flash memory wear and modest computational overhead. UBIFSec is seamlessly added to UBIFS. Cryptographic operations are inserted into UBIFS's existing read/write data path, and the key states are managed by UBIFS's existing index of data nodes.

7.7 Practitioner's Notes

- Our implementation is in software and requires raw access to flash memory. Most user devices, however, use a hardware FTL that obfuscates access to the raw memory. To correctly integrate DNEFS into devices such as SD cards and USB sticks, a DNEFS-enhanced FTL is required.
- The purge period we used was tightly coupled to UBIFS's commit for convenience of implementation. In practice there is no reason for this requirement and there may be better mechanisms possible as the reasons for timely secure deletion are unrelated to the reasons for timely committing.
- UBIFSec did not securely delete metadata as it was handled by a separate component in UBIFS; integrating it would have required more substantial changes. In practice, however, it is always possible to securely delete metadata in the same way as data as it is also stored as pages on the flash memory.
- KSA organization and intelligent caching can greatly improve the read and write latencies and throughput by mitigating the second flash I/O operation.

- The failure of a single KSA erase block may result in a great deal of lost data. A real deployment may find replicating the KSA useful, and perhaps use a secondary flash chip for the purpose.
- We compare wear levelling using a simple economic indicator where each erasure that could have been better levelled is considered equally bad. Depending on the memory's lifetime and the cost of performing wear levelling, other metrics that increase the cost as erase blocks approach their expected lifetime may be more appropriate.
- It is critical that the random numbers used for new keys are taken from a truly random source or, at minimum, a cryptographically suitable random number generator (i.e., a stream cipher's output bytes for a truly random encryption key). Ideally use the same keysize and cryptosystem as the ones used for actual encryption to minimize the risk of data exposure due to a novel cryptanalytic attack. Note that a common source of hardware randomness is hard drive seek times—this is no longer a valid source for flash-based memory.
- Our implementation does not provide authenticity as both the KSA and main storage are the same flash memory; it is therefore assumed that the flash memory handles error correction internally and the storage itself is not malicious. Observe that if an authenticated encryption mode is used, the resulting ciphertext expansion necessitates more storage per data node. This should be stored out-of-band if possible to avoid data nodes becoming page unaligned.

Part III
Secure Deletion for Remote Storage

Chapter 8
Cloud Storage: Background and Related Work

8.1 Introduction

DNEFS uses encryption not as a secrecy technique but instead as a compression technique. This was not, however, the first time encryption was used to facilitate secure deletion. While DNEFS treats the main storage as persistent only for efficiency reasons, a variety of storage media are effectively indelible. Examples of such *persistent* media are write-once media, off-line tape archives, media that leave ample analog remnants [32], and remote storage systems that are outside of the user's control.

In the following chapters, we present a detailed examination of secure deletion solutions for persistent storage. The remainder of this chapter presents different persistent storage media and the object store abstraction, which is a common means by which they are accessed. We then present related work for secure deletion on persistent storage.

Chapters 9, 10, and 11 provide our contributions. Chapter 9 presents our results on generalizing the related work, formalizing the problem of key disclosure using graph theory, and then using the formalism to prove the security of a wide class of potential solutions. Chapter 10 describes a particular solution of our own design, which we then implement and analyze. Chapter 11 considers the problem of an unreliable securely deleting storage media. We design a system robust against failures to delete data, failures to correctly store data, failures to maintain the confidentiality of data, and failures to be available.

8.2 Persistent Storage

As described in Section 3.3, a persistent storage medium implementation is one that is indelible. All data stored on the medium remains stored; any previously stored data is given to the adversary at the time of compromise. While truly limitless per-

J. Reardon, *Secure Data Deletion*, Information Security and Cryptography,
DOI 10.1007/978-3-319-28778-2_8

sistent media are unrealistic, this well models a variety of different settings where secure deletion is not possible: communication over an eavesdropped network, read-only memory, transaction logs such as the bitcoin blockchain [91], paranoia over analog remnants, and failures in executing secure deletion best practices. The storage medium's *persistence* is not a required feature for its correct operation but rather a worst-case scenario.

In some cases, like read-only memory, physical destruction is an option but does not provide effective secure deletion against our adversary who compromises at an unanticipated time.[1] In other cases, like remote storage and network traffic, physical destruction is not an option because the adversary may obtain the data immediately. As a result, we assume that there is no physical destruction option available for the persistent storage, and that all data written to a persistent storage medium is *immediately given to the adversary*. Importantly, we do not consider this to constitute a *compromise*; we still want to securely delete this data.

8.2.1 Securely Deleting and Persistent Combination

While data cannot be deleted from a persistent medium, there exist a variety of mixed-media solutions where it is assumed that the user stores data using both a *persistent* storage medium and a *securely deleting* storage medium [2,3,5,10,13,92, 93]. In these situations, adversarial compromise only concerns the securely deleting storage medium, not the persistent one.

A non-trivializing assumption is that the user is, for some reason, compelled to use the persistent storage medium for storing data; perhaps because the securely deleting medium is small in capacity, slow in performance, inconvenient to use, not able to share data, etc. The securely deleting medium may be, for example, a trusted platform module or a portable smart card that allows users to access remotely stored data anywhere; in both cases we can expect it to only store a limited amount of data. Table 8.1 presents relevant secure-deletion scenarios represented by this model and characterizes example types of media and the reasons why the persistent storage is used.

8.2.2 Cloud Storage

Our model assumes that the cloud storage is untrusted and thus is itself adversarial. While this characterization is not necessarily always true, there is a compelling reason to model it as such: once the data has left the users' control, they can no longer

[1] Recall from Section 4.2.4 that if many units of read-only memory are available and can be independently destroyed, then they can form the constituents of an archive whose erasure granularity is larger than the read and write granularity.

Table 8.1 Situations modelled by persistent storage media.

setting: tape vault
reason to use: cheap massive backups
persistent storage: magnetic tapes
securely deleting storage: guarded machine at tape-drive site
adversary: insider at vault or in transit

setting: remote storage
reason to use: convenience of ubiquitous access
persistent storage: networked file systems
securely deleting storage: smart card, laptop, mobile phone
adversary: operator of remote storage server

setting: forward secrecy
reason to use: shared access to data, communication
persistent storage: network communication
securely deleting storage: session keys, long-term signing key
adversary: network eavesdropper, key compromise

setting: analog remnants
reason to use: limited memory
persistent storage: digital storage
securely deleting storage: human memory
adversary: one with unimaginable forensic capabilities

themselves ensure access control and secure deletion and must instead trust that it is performed correctly.

When data is stored remotely, it may be replicated many times, with backups and snapshots being stored in offline tape vaults. Resources may be shared or security vulnerabilities may enable unauthorized access. Centralized servers become a more-valuable target for attacks. The data may be housed in data centres whose legal jurisdictions differ from those of the user. Legislative requirements may complicate remote storage of some kinds of data, such as the geolocation of banking information. Government adversaries may obtain user data through legal means, for instance, by obtaining surreptitious access to the storage medium through a legal subpoena to the storage provider.

The trustworthiness of an organization can also change over time: bankruptcy may legally require the liquidation of assets to satisfy creditors, or the organization may be purchased by a larger, less trustworthy one. Even private cloud infrastructures are vulnerable to insider attacks, poor configurations, mismanagement, human error, etc.

8.3 Related Work

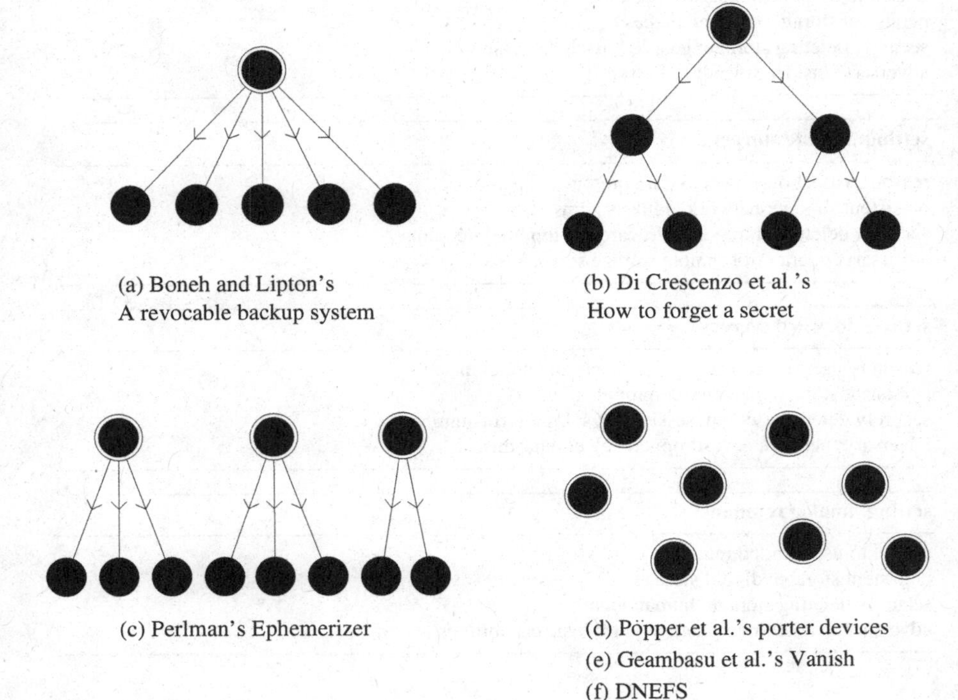

(a) Boneh and Lipton's A revocable backup system

(b) Di Crescenzo et al.'s How to forget a secret

(c) Perlman's Ephemerizer

(d) Pöpper et al.'s porter devices
(e) Geambasu et al.'s Vanish
(f) DNEFS

Fig. 8.1 Key-wrapping structures for secure-deletion solutions for persistent storage. Each node is a key. Directed edges mean that the source node wraps the destination. Circled nodes are keys stored on the securely deleting storage medium. Leaf nodes are used to encrypt data items.

In this section, we describe related work on secure deletion for persistent storage media that are augmented with a securely deleting storage media. We note that some of these works did not explicitly consider this system model; in these cases, it is our opinion that their model fits into this framework.

Some related work uses hierarchical key-wrapping structures to achieve secure deletion. For a visual reference, Figure 8.1 illustrates an example of the general shape of the key structures for relevant related work using directed graphs. In each subfigure, each node (black circle) corresponds to a key; a directed edge means that the source key is used to wrap the destination key. Encircled nodes have no incoming edges and correspond to master keys stored on the securely deleting storage medium; nodes with no outgoing edges are data-item keys used to encrypt individual data items.

Revocable Backup System.

Boneh and Lipton (Figure 8.1 (a)) propose the first scheme that uses secure deletion of cryptographic keys to securely delete encrypted data under computational assumptions [5]. They created a revocable backup system for off-line (i.e., tape) archives consisting of three user-level applications. Backup files are made revocable *before* writing them to tape. Backups are revoked and then securely deleted without needing physical access to the tapes on which they are stored. Each backup is encrypted with a unique key; each key is then encrypted with a temporary master key. Their solution is clocked, so time is discretized into intervals and each interval is assigned a new master key that encrypts all the backup keys. Backups are deleted from the archive simply by *not re-encrypting* the corresponding backup key with the new master key at the next clock edge. They extend their user interface to include master-key management with a *secure-deletion* feature; in their work, they propose to write the new key on paper or on a floppy diskette and then physically destroy the previous one. The paper or floppy diskette therefore constitutes the securely deleting medium while the tape archive constitutes the persistent storage.

How to Forget a Secret.

Di Crescenzo et al. (Figure 8.1 (b)) first explicitly considered secure deletion on a storage medium consisting of both a large *persistent medium* and a small *securely deleting medium* [92]. They divide a fixed-size persistent medium into numbered blocks, which are indexed by a pre-allocated binary tree. The keys to decrypt data are stored in the leaves and the tree's internal nodes store the keys to decrypt the children's keys. The root key is stored in a securely deleting medium. Each change to a data block indexed by the binary tree results in a new key stored in a leaf node and the *rekeying* of all nodes on the path from the leaf to the root. Rekeying means that a new key is generated to encrypt the new key of the children recursively until a new master key is generated and stored on the securely deleting storage. In this scheme, the securely deleting storage medium needs only to store a single key value.

Ephemerizer.

Perlman's Ephemerizer (Figure 8.1 (c)) aims to securely delete communicated messages after an expiration time [2]. Exchanged messages are encrypted using ephemeral keys with a predetermined lifetime. Secure deletion is used to ensure keys are irrecoverable after they expire. Perlman's scheme uses a trusted third party—the Ephemerizer—to manage the ephemeral keys and ensure their irrecoverability after expiration.

Each message is encrypted with a random key, which is then blinded and sent to the Ephemerizer along with the desired message lifetime. The Ephemerizer encrypts the message key with a corresponding ephemeral key based on the desired lifetime. The message encrypted with the random key, along with the random key encrypted with the ephemeral key, are sent as the message. The recipient uses the Ephemerizer, with blinding, to determine the message key. Once the ephemeral key expires, the Ephemerizer no longer possesses it and is therefore unable to decrypt any keys wrapped with it. Thus, the Ephemerizer deletes data at the granularity of an expiration time. Note that this scheme results in an unbounded existential latency as the expiration-time-based keys are created at initialization.

In this scheme, the Ephemerizer is the securely deleting medium: it manages a number of a securely deletable values that are used for many data items in a flat hierarchy, each corresponding to an expiration time. Foreknowledge of the data item's expiration time can prevent costly rewrapping operations that are linear in the number of data items, which is the case for Boneh and Lipton's proposal. The message recipient acts as the persistent storage medium: instead of requiring that recipients perform suitable secure deletion promptly after the message expires, the Ephemerizer is introduced to ensure this is done correctly.

A notable aspect of the system is that the Ephemerizer's operator is not the user but a shared service that is not entrusted to know the contents of messages. Two systems are presented to achieve this: one involving triple encryption and one involving blinded decryption.

File System Design with Assured Delete.

Perlman extends her previous work on the Ephemerizer to efficiently, reliably, and scalably integrate the service into a file system [9]. To improve reliability and availability, she uses multiple Ephemerizers. Secret sharing [94] is used to divide data encryption keys into n shares, any k-sized quorum of which can determine the key. Each share is made securely deletable with a different Ephemerizer. Thus, only k such services need to be available for the key to be available, and only $n-k+1$ such services need to securely delete their key for the key to be securely deleted.

She proposes three manners of storing data with secure deletability: (i) data is stored with its expiration time known in advance; (ii) individual files can be deleted on demand; (iii) classes of files can be created and deleted on demand. The first solution is equivalent to her original Ephemerizer with the addition of a quorum.

The second solution takes Boneh and Lipton's approach [5] and uses Ephemerizers. Each file is encrypted with a unique file key and stored in memory in a file key table; she calls this table the *F-Table*. The encrypted F-Table is periodically backed up to persistent storage—encrypted with a master key that is made securely deletable by a quorum of Ephemerizers. Secure deletion occurs when the corresponding file key is not included in the backup. Perlman notes two risks in this approach. First, there is a high deletion latency depending on the costs of using the Ephemerizer. Second, there is a risk of data loss if the F-Table is surreptitiously corrupted; this risk is because the file keys are continually re-encrypted with new short-lived keys. If the file key is corrupted and then backed up with a new key, the old (correct) version is irrecoverable.

The third solution aggregates files into *classes*, an idea she further develops in collaboration with Tang et al.'s Fade [3]. Instead of providing the lifetime of files, classes ensure that all the files they contain remain available until the Ephemerizer is instructed to delete the class. This addresses both her concerns with her second solution: deletion occurs promptly and class keys remain available from their creation until they are destroyed. If the class key is backed up once in its correct form, then it is available until a quorum of Ephemerizers discards their keys that permit deriving the corresponding shares.

Keeping Data Secret Under Full Compromise Using Porter Devices.

Pöpper et al. [10] (Figure 8.1 (d)) formalize the problem of communicating secretly against an adversary that observes communications between parties at all times and can also perform a coercive attack to compromise both parties' storage media and secret keys or passphrases. They propose a protocol using a trusted porter device, such as a mobile phone, to store and later securely delete keys that encrypt time-limited data.

Messages are encrypted with a session key negotiated by both parties using Diffie-Hellman key negotiation. The storage and timely deletion of session keys is then managed by the porter device. The sender encrypts the message with the session key and then deletes the key. The encrypted message, along with its expiration time, is then sent to the recipient. The recipient retrieves the key from the porter device to decrypt and read the message. When the message's expiration time is reached, the porter device securely deletes the session key that decrypts the message. In this system, the porter device acts as the securely deleting storage medium and the communicating parties' normal storage is the persistent storage.

Fade.

Tang et al.'s Fade [3] extends the third solution from Perlman's File System Design with Assured Delete [9] by explicitly considering cloud storage as the persistent medium and by offering more-expressive deletion policies than expiration dates. An Ephemerizer-like entity acts as the securely deleting medium, but each key that it manages corresponds to a specific *policy atom* that can expire or be revoked. These policy atoms can be combined using logical OR and AND operators, thus allowing more-sophisticated policies to be expressed in a canonical form. The result is a collection of derivable policy keys that can be computed only if the logical expression is TRUE where truth is defined as the securely deleting storage medium storing the corresponding key.

For instance, a policy may state that data is *not securely deleted* if *its expiration time has not elapsed* and *it has not been specifically redacted.* Each conjunct is associated with a key, both of which are needed to decrypt the message. Logical AND is implemented using nested key wrapping, i.e., all keys must be available to derive the corresponding policy key. Logical OR is implemented by having the key corresponding to each OR operand independently wrap the resulting policy key. Similar to the Ephemerizer, Fade deletes data at the granularity of an entire policy.

Policy-Based Secure Deletion.

Cachin et al. also design a policy-based secure deletion system with an expressive policy language [93], as well as cryptographic proofs for all constructions. Their system builds a directed policy graph that maps attributes to policies. Attributes' values are either true or false; boolean values feed forward through the graph. Each node is a threshold operator, e.g., if at least k-out-of-n parents are TRUE then it is TRUE. Logical OR and AND are special cases: $k = 1$ and $k = n$, respectively.

Each attribute is associated with a key, and these keys are stored on a securely deleting storage medium used by the system. When an attribute is no longer true (e.g., a user no longer has a role or the data lifetime has expired), then the key for that attribute is securely deleted. Each node is a threshold-cryptographic operator; if the policy language's expression is no longer true, then the corresponding policy key is no longer retrievable. As with Fade, this solution deletes data at the granularity of an entire policy.

Vanish.

Geambasu et al.'s Vanish (Figure 8.1 (e)) is a system for securely deletable communication over the Internet [13]. Similar to the Ephemerizer, Vanish uses known expiration times when initially storing data. Each communication message is encrypted with a unique key, and this key is divided into shares and stored in a distributed hash table (DHT) [95]. A special *access key* is generated to determine where in the DHT the shares are stored; instead of providing the message recipient with the message's encryption key, the sender instead provides the access key. The access key allows the recipient to determine where to search for the encryption key's shares; once retrieved the key is reconstructed and the message is decrypted.

The security of their scheme relies on the nodes in the DHT together implementing a securely deleting medium. Shares are not intentionally deleted from the DHT but instead gradually disappear due to the DHT's natural churn: the fact that nodes naturally join and leave the system and data must be continually renewed to remain available. Once there are fewer key shares available than the key-sharing threshold, the session key is irrecoverable and so is the message it encrypts.

Vanish imposes no access control on key shares; in fact, key shares are stored on a volunteer network in which anyone can participate and thus learn key material without performing a coercive attack. The authors considered such attacks but presented an economic argument that they are too expensive to succeed. This argument was based on the cost of nodes participating continually in the system. In practice, however, a node may rapidly and repeatedly join, replicate available data, and depart; this can continue until the DHT's entire contents are replicated by the adversary. Follow-up work has shown such low-cost Sybil attacks are feasible in practice [96].

DNEFS.

For completeness, we observe that DNEFS (Figure 8.1 (f)) also provides secure deletion using a small securely deleting key-storage area and a large persistent main storage; they just happen to be the same storage medium (see Chapter 6).

8.4 Summary

Achieving secure deletion on persistent storage media requires encrypting the data and managing the keys on a securely deleting storage medium. Figure 8.1 shows the different key encryption structures of some related work. Boneh and Lipton store a single master key and have linear updates, Di Crescenzo et al. store a single master key and have logarithmic updates [92], Perlman stores multiple master keys that expire at known times in lieu of updates [2], and Vanish [13], Pöpper et al. [10], and DNEFS (Chapter 6) store a linear number of keys with constant updates. Systems that store a linear number of keys are preferred if the securely deleting storage medium can fit them all. Otherwise, the choice of data structure is a trade-off between the costs of reading data versus deleting data and depends on the intended workload.

Di Crescenzo et al.'s binary tree key structure has logarithmic read, write, and delete operations, however it fixes the tree's size and shape before storing any data. Only the values associated with the nodes can change. The total amount of data that can be stored, therefore, is limited to what is initially fixed. Many useful data structures, however, are dynamic: they grow and shrink to accommodate data at the cost of more-complicated update logic. In the next chapter, we prove that any tree-like data structure can be used to provide secure deletion and formalize the requirements on how updates are performed to achieve this. In Chapter 10, we implement a dynamic B-Tree-based securely deleting data structure from the space of data structures we prove secure.

Another concern with all related work is that the securely deleting storage medium is always assumed to have perfect storage properties: it never loses data, it always deletes data, it is always available. These assumptions are unrealistic in practice; it is particularly problematic when the securely deleting storage medium stores a single master key required to access *all data stored*. In Chapter 11, we relax these assumptions and explore the problems that arise as a consequence.

Chapter 9
Secure Data Deletion from Persistent Media

9.1 Introduction

This chapter explores how to securely delete data that is stored on the combination of a persistent storage medium and a securely deleting one, under the assumption that the data cannot be only stored on the securely deleting storage. Instead, the persistent storage stores encrypted versions of all the user's data while the encryption keys required to access it are stored on the securely deleting medium.

To support efficient random-access modifications to data, the data must be encrypted at the appropriate deletion granularity. Small deletion granularities may easily overwhelm the capacity of a limited securely deleting medium. Key wrapping and key derivation are therefore used to build hierarchies of keys where a small number of master keys are used to derive many fine-grained data item keys.

In this chapter, we develop a new approach to reasoning about this problem by modelling adversarial knowledge as a directed graph of keys and verifying the conditions that result in secure data deletion. We define a generic shadowing graph mutation that models how the adversary's knowledge grows over time. We prove that after arbitrary sequences of such mutations one can still securely delete data in a simple and straightforward way. We prove that when using such mutations, data is securely deleted against a computationally bounded unpredictable multiple-access coercive adversary who is additionally given live access to the persistent medium.

The generic shadowing mutation can express the update behaviour for a broad class of dynamic data structures: those whose underlying structure forms a directed tree (henceforth called an arborescence [97]). This includes self-balancing binary search trees and B-Trees [98], but also linked lists and extendible hash tables [99]. It also expresses the update behaviour of the related work presented in Figure 9.4. In the next chapter, we design a B-Tree-based securely deleting data structure from the space of arborescent data structures, implement it, and analyze its performance.

J. Reardon, *Secure Data Deletion*, Information Security and Cryptography,
DOI 10.1007/978-3-319-28778-2_9

9.2 System and Adversarial Model

Our system model is generally consistent with the main model developed in Chapter 3. The user is provided with two storage media: a fixed-sized securely deleting medium and a dynamic-sized persistent storage medium. We assume that the securely deleting medium automatically securely deletes any discarded data, i.e., it behaves like a SECDEL implementation. We also assume that the persistent medium behaves like a PERSISTENT implementation and therefore does not securely delete any data. The goal is to use both these media to provide secure deletion for as many data items as possible.

Our adversary is a computationally bounded unpredictable multiple-access coercive adversary. The adversary also has live access to the persistent storage medium and so learns the data stored on it immediately; adversarial compromise refers only to obtaining access to the securely deleting storage medium. As always, the adversary has full knowledge of the algorithms and implementation of the system of both the persistent and securely deleting media.

For clarity in our presentation, we assume that all keys k have a name $\phi(k) \in \mathbb{Z}^+$, where ϕ is an injective one-way function mapping keys to their name. The key's name $\phi(k)$ reveals no information about the key k—even to an information-theoretic adversary. For example, the key's name could be the current count of the number of random keys generated by the user. We further assume that the adversary can identify the key used to encrypt data through the use of a *name function*, which maps an encrypted block to the corresponding key's name. Hence, given $E_k(\cdot)$, the adversary can compute a name $\phi(k)$. This permits the adversary to organize blocks by their unknown encryption key and recognize if these keys are later known. We do not concern ourselves with the implementation of such a function, but simply empower the adversary to use it.

9.3 Graph Theory Background

The work in this chapter relies heavily on graph theory. For completeness, and to commit to a particular nomenclature, we first briefly review the relevant aspects of graph theory. A more detailed treatment can be found elsewhere [97].

Directed Graphs.

A *directed graph* (henceforth called a *digraph*) is a pair of finite sets (V,E), where $E \subseteq V \times V$. Elements of V are called *vertices* and elements of E are called *edges*. If G is a digraph, then we write $V(G)$ for its vertices and $E(G)$ for its edges.

A digraph's edges are directed. If $(u,v) \in E(G)$, we say the edge goes *from* the *source* u and *to* the *destination* v. The edge is called *outgoing* for u and *incoming* for v. The *indegree* and *outdegree* of a vertex is the number of all incoming and outgoing edges for that vertex. We prohibit self-edges in G: $(u,v) \in V(G) \Rightarrow u \neq v$.

Paths.

A *non-degenerate walk* W of a graph G is a sequence of elements of $E(G)$: $(v_1,u_1),\ldots,(v_n,u_n)$ such that $n \geq 1$ and $\forall i : 1 < i \leq n$, $u_{i-1} = v_i$. The *origin* of W is v_1 and the *terminus* is u_n. We say W *visits* a vertex v (or equivalently, v is on W) if W contains an edge (v,u) or v is the terminus. A *non-degenerate path* P is a non-degenerate walk such that no vertex is visited more than once. Additionally, a graph with n vertices has n *degenerate paths*—zero-length paths that visit no edges and whose origin and terminus are $v \in V(G)$. A *cycle* is a non-degenerate walk C whose origin equals its terminus, and all other vertices on the walk are visited once. A *directed acyclic graph* is a digraph with no cycles.

A vertex v is *reachable* from vertex u if there is a directed path from u to v. If there is only one such path, then we say that v is uniquely reachable from u and use P_v^u to denote this path. The *ancestors* of a vertex v, called $\mathrm{anc}_G(v)$, is the largest subset of $V(G)$ such that v is reachable from each element. The *descendants* of a vertex u, called $\mathrm{desc}_G(u)$, is the largest subset of $V(G)$ such that each element is reachable from u. If P_v^u is a directed path from u to v, then u is an *ancestor* of v and v is a *descendant* of u. Because of degenerate paths, all vertices are their own ancestors and descendants.

Subdigraphs.

A *subdigraph* S of a digraph G is a digraph whose vertices are a subset of G and whose edges are a subset of the edges of G with endpoints in S. Formally, a subdigraph has vertices $V(S) \subseteq V(G)$ and edges $E(S) \subseteq E(G)|_{V(S)\times V(S)}$. A subdigraph is called *full* if $E(S) = E(G)|_{V(S)\times V(S)}$. A *subdigraph induced by a vertex* v, denoted G_v, is a full subdigraph whose vertices are v and all vertices are reachable from v in G. Formally, $V(G_v) = \mathrm{desc}_G(v)$ and $E(G_v) = E(G)|_{V(G_v)\times V(G_v)}$.

Arborescences and Mangroves.

An *arborescence* A *diverging from a vertex* $r \in V(A)$ is a directed acyclic graph A whose edges are all directed away from r and whose underlying graph (i.e., the undirected graph generated by removing the direction of A's edges) is a (graph-theoretic) tree [97]. The vertex r is called the *root* and it is the only vertex in A that has no incoming edges; all other vertices have exactly one incoming edge (Theorem VI.1 [97]). There is no non-degenerate path in A with r as the terminus, and for all other vertices $v \in V(A)$ there is a unique path P_v^r (Theorem VI.8 [97]). To show that a graph A is an arborescence, it is necessary and sufficient to show that A has the following three properties (Theorem VI.26 [97]): (i) A is acyclic, (ii) r has indegree 0, (iii) $\forall v \in V(A), v \neq r \Rightarrow v$ has indegree 1.

A directed graph is a *mangrove* if and only if the subdigraph induced by every vertex is an arborescence. This means that, for every pair of vertices, either one is uniquely and unreciprocatedly reachable from the other or neither one is reachable from the other. Observe that an arborescence is also a mangrove, as all its vertices induce arborescences. Figure 9.1 shows an example mangrove as well as an arborescent subdigraph induced by a vertex.

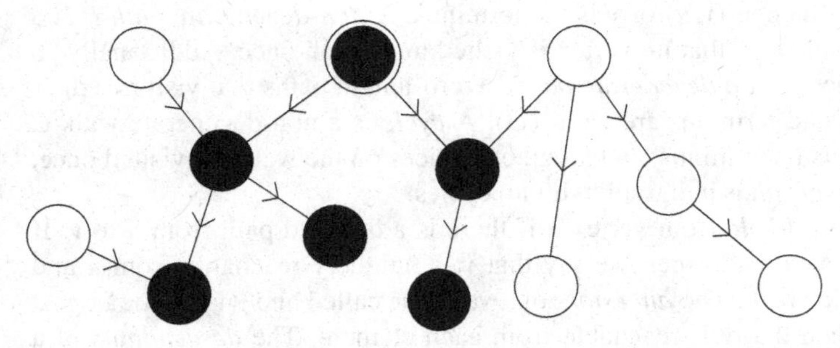

Fig. 9.1 An example mangrove. Shaded vertices belong to the arborescent subdigraph induced by the circled vertex.

9.4 Graph-Theoretic Model of Key Disclosure

We now characterize secure deletion in the context of key wrapping and persistent storage. We use this to prove the security of a broad class of mutable data structures when used to retrieve and securely delete data stored on persistent storage. First, we define a key disclosure graph and show how it models adversarial knowledge. We then prove graph-theoretic conditions under which data is securely deleted against our worst-case adversary. Finally, we define a generic shadowing graph mutation and prove that all valid instantiations of the mutation's parameters preserve a graph property that simplifies secure deletion.

9.4.1 Key Disclosure Graph

In this section, we characterize the information obtained by the adversary and describe a way to structure it. We begin by limiting the functions the user computes on encryption keys to *wrapping* and *hashing*. Wrapping means that a key k is symmetric-key encrypted with another key k' to create $E_{k'}(k)$. With k' and $E_{k'}(k)$ one can compute k, while $E_{k'}(k)$ alone reveals no information about k to a computationally bounded entity. Hashing means that a key k can be used to compute a one-way function $H(k)$ such that $H(k)$ reveals no information about k to a computationally bounded entity. Furthermore, we require that no plain-text data is ever written onto the persistent medium.

The process of generating keys and using keys to wrap other keys induces a directed graph: nodes correspond to encryption keys and directed edges correspond to the destination key being wrapped by the source key. Knowledge of one key gives access to the data encrypted with it as well as any keys corresponding to its vertex's destinations. Recursively, all keys corresponding to descendants of a vertex are computable when the key corresponding to the ancestor vertex is known. In other words, if one knows the key associated with the origin of a path in this graph, one can compute the key associated with the terminus. We call this graph the *key disclosure graph*, whose definition follows.

Definition 9.1. Given a set K of encryption keys generated by the user, an injective one-way vertex naming function $\phi : K \rightarrow \mathbb{Z}^+$, and a set of wrapped keys C, then the *key disclosure graph* is a directed graph G constructed as follows: $\phi(k) \in V(G) \Leftrightarrow k \in K$ and $(\phi(k), \phi(k')) \in E(G) \Leftrightarrow E_k(k') \in C$.

The user can construct and maintain such a key disclosure graph by adding nodes and edges when performing key generation and wrapping operations, respectively. The adversary can also construct this graph using its name function: whenever ciphertext is given to the adversary, the name corresponding to its encryption key is computed and added as a vertex to the graph with the ciphertext stored alongside. The adversary may only learn some parts of the key disclosure graph; we use $G^{adv} \subseteq G$ to represent the subgraph known to the adversary. For instance, the client may not write all the wrapped key values it computes to the persistent storage, or the adversary may not be able to read all data in the persistent storage. In the worst case, however, the adversary gets all wrapped keys and so $G^{adv} = G$; it is this worst case for which we prove our security.

If the adversary later learns an encryption key (e.g., through compromise), then the key's corresponding ciphertext can be decrypted. If the plaintext contains other encryption keys, then the adversary can determine the names of these keys to determine the edges directed away from this vertex. Therefore, the adversary can follow paths in G^{adv} starting from any vertex whose corresponding key it knows, thus deriving unknown keys.

The adversary's ability to follow paths in the key disclosure graph is independent of the age of the nodes and edges. In our scenario and adversarial model, every time data is stored on the persistent medium, the key disclosure graph G—and possibly the adversary's key disclosure graph G^{adv}—grows. After learning a key, the adversary learns all paths originating from the corresponding vertex in G^{adv}. The keys corresponding to vertices descendant to that origin are then known to the adversary along with the data they encrypt. Therefore, the user must perform secure deletion while reasoning about the adversary's key disclosure graph. Moreover, if the user is unaware of the exact value of $G^{adv} \subseteq G$, then they must reason about $G^{adv} = G$.

9.4.2 Secure Deletion

Secure data deletion against an adversary with live access to the persistent storage medium requires that the data's encryption key is securely deleted as well as any values that may derive its value. This means that any ancestor of the data's corresponding vertex in the adversary's key disclosure graph must be securely deleted. This is because a vertex v is reachable from another vertex u in the key disclosure graph if and only if $\phi^{-1}(v)$ is computable from $\phi^{-1}(u)$. Definition 9.2 now defines secure deletion in terms of paths in the key disclosure graph.

Definition 9.2. Let $G = (V, E)$ be the key disclosure graph for a vertex naming function ϕ, a set of keys K, and a set of ciphertexts C, and let $G^{adv} \subseteq G$ be the adversary's subdigraph of the key disclosure graph. Let $R = \{r_1, \ldots, r_n\} \subseteq K$ be the set of keys stored by the user in the securely deleting medium. Let D be data stored on the persistent medium encrypted with a key $k \in K$. Let $R_{\text{live}} \subseteq R$ be the set of keys stored in the securely deleting medium at all times when D is alive (i.e., the times between the data's creation and deletion events).

Then D is *securely deleted against a computationally bounded coercive adversary* provided that no compromise of the securely deleting medium occurs when it stores an element of R_{live}, and for all $r \in R \setminus R_{\text{live}}$, there is no path in G^{adv} from $\phi(r)$ to $\phi(k)$.

This definition reflects the following facts: (i) a computationally bounded adversary cannot recover the data D without the key k, (ii) the only way to obtain k is through compromise or through key unwrapping, (iii) an adversary that compromises at all permissible times only obtains $R \setminus R_{live}$ directly and $\bigcup_{r \in R \setminus R_{live}} \text{desc}(\phi(r))$ through unwrapping, and (iv) k is not within this set.

Observe that this definition requires that no compromise occurs during which time the securely deleting medium stores an element of R_{live}—the set of keys stored in the securely deleting storage medium during the lifetime of the data being securely deleted. This is larger than or equal to the data's lifetime, e.g., by extending in both directions to the nearest commit event.

We have shown that to securely delete the data corresponding to a vertex v, we must securely delete data corresponding to all ancestors of v that are not already securely deleted. This is burdensome if it requires a full graph traversal, because the adversary's key disclosure graph perpetually grows. We make this efficient by establishing an invariant of the adversary's key disclosure graph: there is at most one path between every pair of vertices (i.e., the graph is a mangrove). In the next section, we define a family of graph mutations that preserves this invariant.

9.5 Shadowing Graph Mutations

Shadowing is a concept in data structures where updates to elements do not occur in-place. Instead, a new copy of the element is made and references in its parent are updated to reflect this [100]. This results in a new copy of the parent, propagating shadowing to the head of the data structure. We now define a generalized graph mutation, called a *shadowing graph mutation*, and show that if any shadowing graph mutation is applied to a mangrove, then the resulting *mutated graph* is also a mangrove. The mangrove property is therefore maintained throughout all possible histories of shadowing graph mutations.

Mangroves have at most one possible path between every pair of vertices. This simplifies secure deletion of data, as illustrated in Figure 9.2. Computing the set of all ancestors of a vertex—those vertices that must be also securely deleted—is done by taking the union of the unique paths to that vertex from each of the vertices whose corresponding keys are locally stored by the user. Determining the *unique* tree path to find data is easily done by overlaying a search-tree data structure (e.g., a B-Tree). Moreover, if the user only stores one local key at any time, taking care to securely delete old keys, then data can be securely deleted by just securely deleting the vertices on a single path in the key disclosure graph.

Figure 9.2 shows an example mutation, where the old key disclosure graph G is combined with G_S and the edges $\hat{e}_1, \hat{e}_2$ to form the new key disclosure graph G'. The new nodes and edges correspond to the user generating new random keys and sending wrapped keys to the adversary, respectively. The node r represents the user's current stored secret key; the shaded nodes are r's descendants—those nodes whose corresponding keys are computable by the user. In the resulting graph G', we see that r' corresponds to the new user secret, resulting in a different set of shaded descendant vertices. In particular, the mutation securely deleted the leaves l_2 and l_4 while adding new leaves l_6 and l_7.

To perform the mutation, the user prepares T—a graph that contains the vertices to shadow. In the post-mutated graph G', no vertex in T is reachable from any vertex in G_S. The only vertices in G that are given a new incoming edge from a vertex in G_S are those in the set $W(G,T)$: vertices outside T that have an incoming edge from a vertex in T. Formally, if G is a mangrove, $r \in V(G)$, and T is an arborescent subdigraph of G_r diverging from r, then $W(G,T) = \{v \in V(G) \setminus V(T) | \exists\, x \in V(T) \,.\, (x,v) \in E(G)\}$.

To ensure that G' is a mangrove, we must constrain the edges that connect G_S to G. We require that any connecting edge $\hat{e}$ goes from G_S to $W(G,T)$ and that each vertex in $W(G,T)$ receives at most one such incoming edge.

Mangroves ensure that during the entire course of operations, no additional paths to compute keys were ever unexpectedly generated. Therefore, the client-side cost of managing the key disclosure graph is significantly reduced; secure deletion is achieved by shadowing along the unique path from the vertex that should be deleted to the root vertex, and securely deleting the key corresponding to the previous root.

Formally, a tuple $(G, r, G_S, T, \hat{E})$ is a *shadowing graph mutation* if it has the following properties:

- G is a mangrove, called the *pre-mutated graph*.
- r is a vertex of G.
- G_S is an arborescence diverging from $r_S \in V(G_S)$ such that $V(G) \cap V(G_S) = \emptyset$. It is called the *shadow graph*.

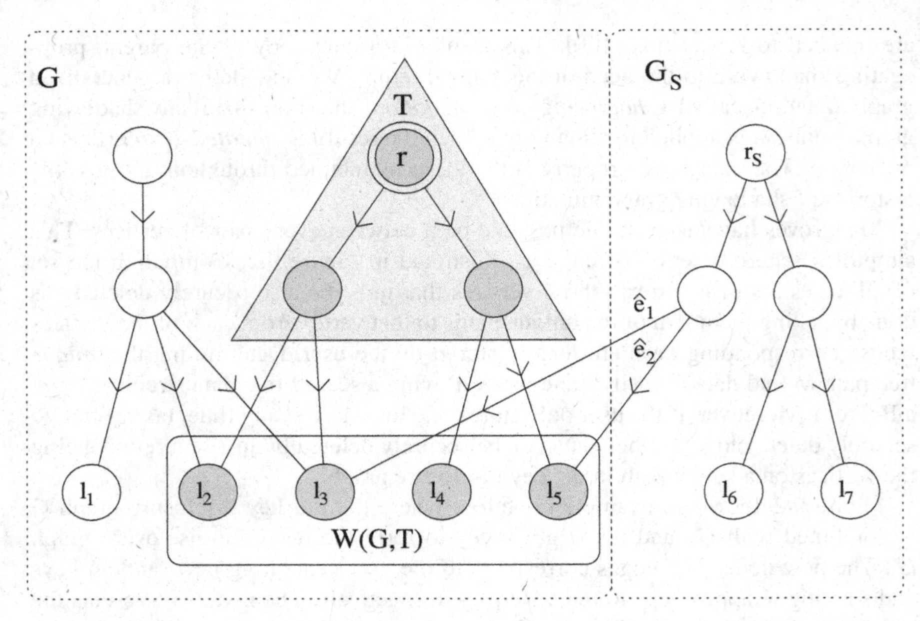

(a) mutation parameters

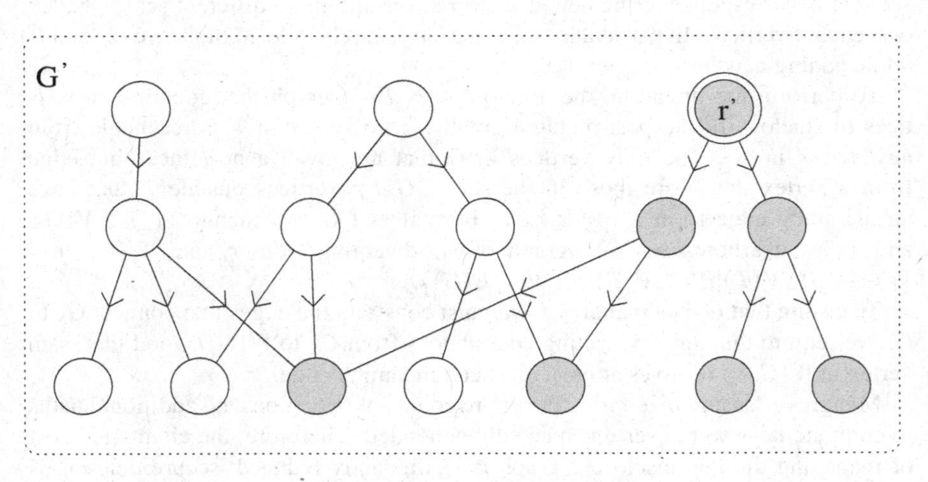

(b) post–mutated graph

Fig. 9.2 An example of a shadowing mutation. (a) The parameters of a shadowing graph mutation. (b) The resulting graph. The pre-mutated graph G is combined with the shadowing graph G_S and connecting edges $\hat{E} = \{\hat{e}_1, \hat{e}_2\}$ to form G'. Shaded vertices are the vertices reachable from the circled vertex.

- T is a subdigraph of G_r such that T is an arborescence diverging from r. It is called the *shadowed graph*.
- $\hat{E}$ is a set of directed edges such that
 (i) $\forall (i,j) \in \hat{E} \,.\, i \in V(G_S) \wedge j \in W(G,T)$ and
 (ii) $\forall \{(i,j),(i',j')\} \subseteq \hat{E} \,.\, i \neq i' \Rightarrow j \neq j'$ (i.e., $\hat{E}$ is injective).

A graph mutation contains the initial graph along with the parameters of the mutation. We assume there exists a function μ that takes as input a graph mutation $(G,r,G_S,T,\hat{E})$ and outputs the mutated graph G', defined by $V(G') = V(G) \cup V(G_S)$ and $E(G') = E(G) \cup E(G_S) \cup \hat{E}$. Observe that the sets in the unions are all disjoint. Moreover, every resulting path in G' has one of the following forms: P, P_S, or $(P_S, \hat{e}, P)$, where P is a path visiting only vertices in $V(G)$, P_S is a path visiting only vertices in $V(G_S)$, and $\hat{e} \in \hat{E}$.

9.5.1 Mangrove Preservation

To simplify the enumeration of a vertex's ancestors in the key disclosure graph, which must be securely deleted in order to delete that vertex, we require as an invariant that the key disclosure graph is always a mangrove. We establish this by showing that, given a shadowing graph mutation, the mutated graph is always a mangrove. Since the graph with a single vertex is a mangrove, all sequences of shadowing mutations beginning from this mangrove preserve this property.

Lemma 9.1. *Let G be a mangrove, $r \in V(G)$, and T an arborescent subdigraph of G_r diverging from r. Then $\forall i,j \in W(G,T), i \neq j \Rightarrow desc_G(i) \cap desc_G(j) = \emptyset$.*

Proof. We prove the contrapositive. Suppose that $v \in \mathrm{desc}_G(i) \cap \mathrm{desc}_G(j)$. Then there exist distinct paths P^i_v and P^j_v. Since $i,j \in V(G_r)$, there exist distinct paths P^r_i and P^r_j. Consequently, $P^r_i P^i_v$ and $P^r_j P^j_v$ are two paths from r to v in G_r. Since G_r is an arborescence, these two paths must be equal and so (without loss of generality) $P^r_v = P^r_i P^i_j P^j_v$ and $P^r_j = P^r_i P^i_j$. However, by definition of $W(G,T)$, all edges except the final one in P^r_i and P^r_j are in $E(T)$. If P^i_j is non-degenerate, then $P^r_i P^i_j \neq P^r_j$ as P^r_i has an edge outside of T followed by more than one edge. Therefore, P^i_j is degenerate and $i = j$, as needed.

Lemma 9.2. *If $(G,r,G_S,T,\hat{E})$ is a valid shadowing graph mutation and $G' = \mu(G,r,G_S,T,\hat{E})$, then G' is acyclic.*

Proof. Since the mutation is valid, G is a mangrove. Suppose to the contrary that G' has a cycle C. By construction of $V(G')$, there are three cases:
(i) All of C's vertices are in $V(G)$. Then C is a cycle in G, which contradicts G being a mangrove.
(ii) All of C's vertices are in $V(G_S)$. Then C is a cycle in G_S, which contradicts G_S being an arborescence.

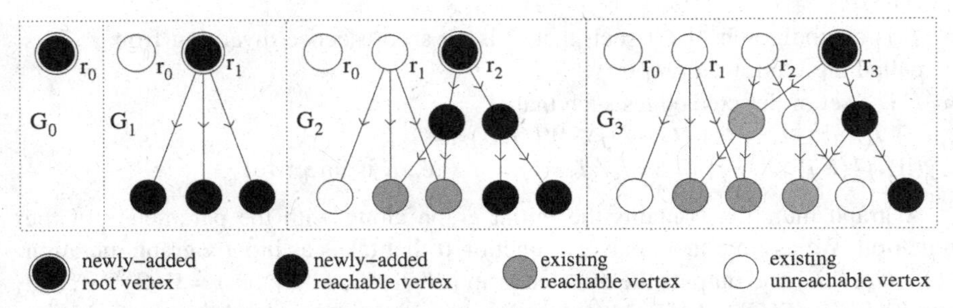

Fig. 9.3 Example key disclosure graph evolving due to a shadowing graph mutation chain. All graphs except G_0 result from applying a shadowing graph mutation on the previous graph. Black nodes are ones added by the most recent mutation with the root node circled; grey nodes are ones from the previous graph that are still reachable from the new root; white nodes are ones from the previous graph that are no longer reachable.

(iii) C's vertices are a mixture of vertices from $V(G)$ and $V(G_S)$. Suppose C visits $v \in V(G)$ and $u \in V(G_S)$. Then C can be divided into two paths $C = P_u^v P_v^u$, but no such path P_u^v exists.

Theorem 9.1. *If* $(G, r, G_S, T, \hat{E})$ *is a valid shadowing graph mutation and* $G' = \mu(G, r, G_S, T, \hat{E})$, *then* G' *is a mangrove.*

Proof. By the definition of a mangrove, we must show that all vertices in G' induce arborescences. Suppose to the contrary that there is some $r \in V(G')$ such that G'_r is not an arborescence. Then (at least) one of the three necessary and sufficient conditions of an arborescent graph is violated:
(i) G'_r is not acyclic. This implies that G' is not acyclic, which contradicts Lemma 9.2.
(ii) The indegree of $r \neq 0$. Then r must have at least one incoming edge, from a vertex v. This results in a cycle, since v is reachable from r by construction of the induced graph G'_r, also contradicting Lemma 9.2.
(iii) There is some $v \in V(G'_r)$ such that $v \neq r$ and indegree of $v \neq 1$.

As the first two conditions lead to immediate contradictions, we assume that the final condition is violated. Moreover, since v is a vertex on an induced graph, there is a path from r to v and thus v must have at least one incoming edge and therefore the indegree of $v \geq 2$. By the induced graph G'_v's construction, both parents of v are reachable from r, and so there are two distinct paths P_v^r and Q_v^r in G' from r to v. We have two cases: either $r \in V(G)$ or $r \in V(G_S)$.

Suppose that $r \in V(G)$, and so all vertices of P_v^r and Q_v^r are elements of $V(G)$. Also, by construction, $E(G')|_{V(G)\times V(G)} = E(G)$, and thus all edges of P_v^r and Q_v^r are elements of $E(G)$. Therefore, P_v^r and Q_v^r are distinct paths from r to v in G, contradicting G being a mangrove.

Now suppose that $r \in V(G_S)$. If $v \in V(G_S)$, then P_v^r and Q_v^r are distinct paths entirely in G_S, which contradicts G_S being an arborescence. So $r \in V(G_S)$ and $v \in V(G)$. We decompose the paths as follows: $P_v^r = P_u^r, (u, w), P_v^w$ and $Q_x^r, (x, y), Q_v^y$, where (u, w) and (x, y) are elements of $\hat{E}$. We know that $P_v^r \neq Q_v^r$, and so there are

four different cases based on the edge in $\hat{E}$:
(i) If $(u,w) = (x,y)$, i.e., both paths cross from G_S to G over the same edge in $\hat{E}$, then the two paths must differ elsewhere. Either $P_u^r \neq Q_x^r$ or $P_v^w \neq Q_v^w$. As we have seen before, however, this contradicts either G_S being an arborescence or G being a mangrove, respectively.
(ii) If $u \neq x$ and $w = y$, then (u,w) and (x,w) are distinct edges in $\hat{E}$, a violation of its construction. This contradicts $(G,r,G_S,T,\hat{E})$ being a valid shadowing mutation.
(iii), (iv) If $w \neq y$ then we have distinct paths P_v^w and Q_v^y in G. Since both paths terminate at the same vertex, either w or y is the ancestor of one of the other's descendants. This contradicts Lemma 9.1.

In conclusion, such distinct paths P_v^r and Q_v^r cannot exist. Therefore, for all $r \in V(G')$, G'_r is an arborescence and so G' is a mangrove.

9.5.2 Shadowing Graph Mutation Chains

Definition 9.2 tells us that we can achieve secure deletion with appropriate constraints on the shape of the key disclosure graph. We now show that performing a natural sequence of shadowing graph mutations satisfies these constraints, effecting simple secure deletion.

Definition 9.3. A sequence of shadowing graph mutations $\mathscr{M} = (M_0, \ldots, M_p)$, where each $M_i = (G_i, r_i, G_{S,i}, T_i, \hat{E}_i)$, is a *shadowing graph mutation chain* if (i) $G_0 = (\{\phi(0)\}, \emptyset)$, (ii) $r_0 = \phi(0)$, (iii) $\forall i > 0 \,.\, G_i = \mu(M_{i-1})$, and (iv) $\forall i > 0 \,.\, r_i = r_{S,i-1}$.

A shadowing graph mutation chain describes a sequence of mutations applied on a key disclosure graph. Figure 9.3 shows an example key disclosure graph evolution as the result of three mutations. Each mutation in the chain is applied on the graph that results from the previous mutation, except for the base case. Observe that r_i—the root vertex in T_i—is always $r_{S,i-1}$, the root vertex added by the shadowing graph in the previous mutation (or the 'zero' key for the base of recursion).

We now prove our main result about the interplay of secure deletion and shadowing graph mutation chains. For convenience, given $M = (G,r,G_S,T,\hat{E})$, we say that a vertex $v \in V(G)$ is reachable in M if there exists a path from r to v in G.

Lemma 9.3. *Let $\mathscr{M} = (M_0, \ldots, M_p)$ be a shadowing graph mutation chain. Any vertex v first reachable in M_i and last reachable in M_{i+k} $(k \geq 0)$ is reachable in all intermediate mutations $M_{i+1}, \ldots, M_{i+k-1}$.*

Proof. Suppose to the contrary that there exists a j, $i < j < i+k$, such that v is not reachable in M_j. By construction of shadowing graph mutations, $v \in V(G_i) \Rightarrow v \in V(G_j) \Rightarrow v \notin V(G_{S,j})$. Select the largest such j, so that v is reachable in M_{j+1}, and so there exists a path $P_v^{r_{j+1}}$ in G_{j+1}. Since $r_{j+1} \in V(G_{S,j})$ and $v \notin V(G_{S,j})$, such a path has the form $P_{\hat{e}}^{r_{j+1}}(\hat{e},\hat{e}')P_v^{\hat{e}'}$ where $(\hat{e},\hat{e}') \in \hat{E}_j$ and $P_v^{\hat{e}'}$ is a path in G_j. Then $\hat{e}' \in W(G_j,T_j)$ and so $P_{\hat{e}'}^{r_j}$ is a path in G_j, implying that $P_{\hat{e}'}^{r_j}P_v^{\hat{e}'}$ is a path from r_j to v in G_j, which leads to a contradiction.

Lemma 9.3 tells us that, when building shadowing graph mutation chains as described, once some reachable vertex becomes unreachable then it remains permanently unreachable. Secure deletion is achieved by a single mutation that makes the corresponding vertex unreachable from the new root. We now prove our final result on achieving secure deletion with shadowing graph mutation chains.

Theorem 9.2. *Let $\mathcal{M} = (M_0, \ldots, M_p)$ be a shadowing graph mutation chain with resulting key disclosure graph $G = \mu(M_p)$. Let $T = (t_0, \ldots, t_p)$ be the (strictly increasing) sequence of timestamps such that at time t_i*
(i) $\mu(M_i)$ is performed,
(ii) the value $k_{i+1} = \phi^{-1}(r_{i+1})$ is stored in the securely deleting memory, and
(iii) all previous values stored are securely deleted.
Let D be data encrypted with the key k whose corresponding vertex $v = \phi(k)$ is reachable only in $M_i, \ldots, M_{i+l}$. Then D's lifetime is bounded by t_i and t_{i+l}, and D is securely deleted provided no compromise occurs during this time.

Proof. The proof is by establishing the premises required for Definition 9.2. First, $R = \{k_0, \ldots, k_p\}$ and $R_{\text{live}} = \{k_i, \ldots, k_{i+l}\}$ which means that $R_{\text{dead}} = \{k_0, \ldots, k_{i-1}\} \cup \{k_{i+l+1}, \ldots, k_p\}$. Because no compromise occurs from time t_i until t_{i+l}, to apply Definition 9.2 we must only show that for all $k_j \in R_{\text{dead}}$, there is no path from $\phi(k_j)$ to v in $G = \mu(M_p)$.

Assume to the contrary that there is a $k_j = \phi^{-1}(r_j) \in R_{\text{dead}}$ such that there is a path $P_v^{r_j}$ in $G = \mu(M_p)$. Since v is unreachable in M_j, $P_v^{r_j}$ is not a path in G_j. So there must be an edge (u, v) on $P_v^{r_j}$ such that $u \in V(G_j), (u, v) \in E(G)$, and $(u, v) \notin E(G_j)$. Then $\exists m \geq 0 : (u, v) \notin E(G_{j+m}) \wedge (u, v) \in E(\mu(M_{j+m}))$, that is, some mutation adds (u, v) to the key disclosure graph. By construction, $E(\mu(M_{j+m})) = E(G_{j+m}) \cup E(G_{S,j+m}) \cup \hat{E}_{j+m}$, and since $u \notin V(G_{S,j+m}) \Rightarrow (u, v) \in E(G_{j+m})$, a contradiction. Definition 9.2 therefore tells us that D is securely deleted.

Theorem 9.2 shows that mangrove-shaped key disclosure graphs that evolve through a sequence of shadowing graph mutations provide simple criteria for secure data deletion. Interestingly, as we show next, some related work also induce mangrove-shaped key disclosure graphs.

9.5.3 Mangrove Key Disclosure Graphs in Related Work

Figure 8.1 in Section 8.3 illustrated the key wrapping structures for some related work. Figure 9.4 extends this earlier figure to show example key disclosure graphs that can be generated after a couple update operations for each of the related work. Observe that they all have mangrove-shaped key disclosure graphs. As a result, the security of these systems follows from the results in this section. Admittedly, the machinery of our proofs is beyond what is necessary if our only goal is to prove the security of a clocked keystore such as DNEFS. Nevertheless, we observe that its security conveniently follows as a corollary from this work due to the fact that its key disclosure graph is a mangrove.

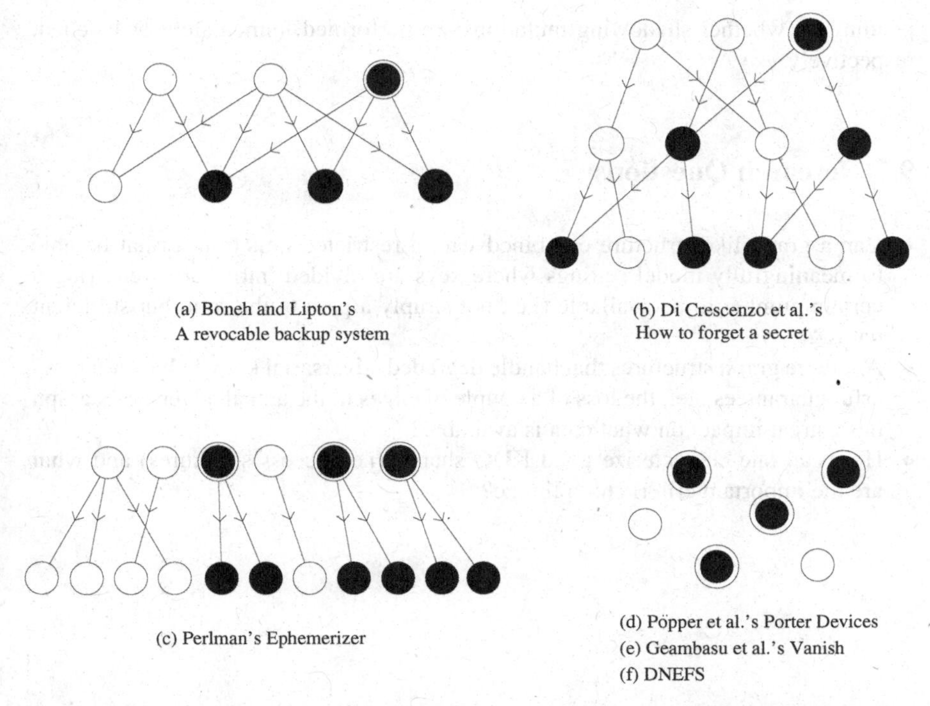

Fig. 9.4 Mangrove-shaped key disclosure graphs for related work. Circled nodes represent keys currently stored on the securely deleting storage; black nodes are currently derivable keys; white nodes are securely deleted keys.

9.6 Summary

We developed a general approach for the design and analysis of secure-deletion solutions from persistent media. We defined a key disclosure graph that models the growth of adversarial knowledge as wrapped keys are written to the persistent storage. We defined the conditions by which data is securely deleted against this adversary. We showed that if the key disclosure graph has the shape of a mangrove then ensuring secure deletion is more easily achieved: no additional data must be stored about the adversary's knowledge other than the data structure used to manage currently valid data.

To ensure that the key disclosure graph retains its mangrove property, we defined a shadowing graph mutation, which is sufficiently generic to express the update behaviour of any arborescent data structure. We proved that applying any shadowing graph mutation to any mangrove always results in a mangrove, and that chains of these mutations can be constructed to reflect arbitrary sequences of the data structure storing and deleting data items. This provides secure data deletion against a computationally bounded unpredictable multiple-access coercive adversary, turning the storage medium into either a SECDEL or SECDEL-CLOCK implementation, depending on whether shadowing mutations are performed immediately or batched, respectively.

9.7 Research Questions

- Can a graph-like structure combined with a restricted mutation format be able to meaningfully model settings where keys are divided into shares requiring a certain number to be available (i.e., not simply a path to the root, but sufficient paths)?
- Are there graph structures that handle degraded adversarial knowledge with powerful guarantees, i.e., the loss of a couple of edges in the actual adversary's graph has a great impact on what data is available?
- How can one characterize good KDG shapes (i.e., access structures) and what are the important criteria to optimize?

Chapter 10
B-Tree-Based Secure Deletion

10.1 Introduction

This chapter builds on the previous chapter of secure deletion for persistent storage by using a small securely deleting storage medium, which explored the general space of possibilities. We design and implement a concrete securely deleting data structure from this space. Our motivation is to provide a *dynamic* data structure, whose capacity can grow and shrink as necessary based on the current requirements.

The B-Tree implements a securely deleting key-value map that maps data handles to data items; new pairs can be inserted, existing pairs can be removed, and any stored data item can be updated. Our B-Tree collects multiple updates and performs them in batch. It therefore consists of two parts, a skeleton tree managed locally to the user and the full tree stored on the persistent storage medium. Periodically, all changes local to the skeleton tree are collected into a single shadowing graph mutation and applied to the full tree. After each update, a new root key value is stored in the securely deleting storage medium, which divides time into deletion epochs; the solution then behaves like a securely deleting clocked implementation with a corresponding deletion latency (Figure 3.2 in Chapter 3). An optional crash-safety mechanism we propose further adds existential latency (Figure 3.3 in Chapter 3).

We implement our B-Tree-based instance and test it in practice. Our implementation offers a virtual block device interface, i.e., it mimics the behaviour of a typical hard drive. This permits any block-based file system to use the device as a virtual medium, and so any medium capable of storing and retrieving data blocks can therefore be used as the persistent storage. We show that our solution achieves secure deletion from persistent media without imposing substantial overhead through increased storage space or communication. We validate this claim by implementing our solution and analyzing its resulting overhead and performance. We examine our design's overhead and B-Tree properties for different caching strategies, block sizes, and file system workloads generated by `filebench` [101]. We show that the caching strategy approximates the theoretical optimal (i.e., Bélády's "clairvoyant"

J. Reardon, *Secure Data Deletion*, Information Security and Cryptography,
DOI 10.1007/978-3-319-28778-2_10

strategy [102]) for many workloads and that the storage and communication costs are typically only a small percentage of the cost to store and retrieve the data itself.

10.2 System and Adversarial Model

This chapter focuses on the design and implementation of a solution whose general space and security is described in Chapter 9. The system and adversarial model we use is identical to the one described in Section 9.2. The update behaviour of our B-Tree design is expressible as a shadowing graph mutation. By applying the results of Chapter 9, our B-Tree comes from the general space of possible solutions that we proved secure.

10.3 Background

A B-Tree is a self-balancing search tree [98] that implements a key-value map interface. B-Trees are ubiquitously deployed in databases and file systems as they are well-suited to accessing data stored on block devices—devices that impose some non-trivial minimum I/O size.

A B-Tree of order N is a tree where each node has between $\lceil \frac{N}{2} \rceil$ and N child nodes, and every leaf has equal depth [98]. (The root is exceptional as it may have fewer than $\lceil \frac{N}{2} \rceil$ nodes.) The order of a B-Tree node is chosen to fit perfectly into a disk block, which maximizes the benefit of high-latency disk operations that return at minimum a full block of data. B-Trees typically store search keys whose corresponding values are stored elsewhere; leaf nodes store the location where the data can be found. The basic key-mutating operations are `add`, `modify`, and `remove`. Because adding and removing children may violate the balance of children in a node, `rebalance`, `fuse`, and `split` are used to maintain the tree balance property.

10.3.1 B-Tree Storage Operations

The `add`, `modify`, and `remove` functions begin with a `lookup` function, which takes a search key and follows a path in the tree from the root node to the leaf node where the search key should be stored. `Add` stores the search key and a reference to the data in the leaf node. `Modify` finds where the data is stored and replaces it with new data; alternatively it can store the new version out-of-place and update the reference. `Remove` removes the reference to data in the leaf node. Both `add` and `remove` change the number of children in a leaf node, which can violate the balance property.

10.3.2 B-Tree Balance Operations

A B-Tree of order N is balanced when (i) the number of children of each non-root node is inclusively between $\lceil \frac{N}{2} \rceil$ and N and (ii) the number of children in the root node is less than or equal to N. When there are more or fewer children than these thresholds, the node is overfull or underfull, respectively, and must be balanced.

Overfull nodes are `split` into two halves and become siblings. This requires an additional index in their parent, which may in turn cause the parent to become overfull. If the root becomes overfull, then a new root is created; this is the only way the height of a B-Tree increases.

Underfull nodes can be either `rebalanced` or `fused` to restore the tree balance property. Rebalancing takes excess children from one of the underfull node's siblings; this causes the parent to reindex the underfull node and its generous sibling and afterwards neither node violates the balance properties. If both the node's siblings have no excess children, then the node is fused with one of its siblings. This means that the sibling is removed and its children are given to the underfull node. This removes one child from its parent, which can cause the parent to become underfull and can propagate to the root. The root node is uniquely allowed to be underfull. If, however, after a fuse operation the root has only one child, then the root is removed and its sole child becomes the new root. This is the only way the height of a B-Tree decreases.

10.4 Securely Deleting B-Tree Design

We use a B-Tree to organize, access, and securely delete data. We assume that only a constant number of B-Tree nodes can be stored on the securely deleting storage medium. Consequently, both data and B-Tree nodes are stored on the persistent storage medium, and they are encrypted before being stored.

Data blocks are encrypted with a random key. The index for the data block, along with its encryption key, is then stored as a leaf node in the B-Tree. The nodes themselves are encrypted with a random key and stored on the persistent medium. Inner nodes of the B-Tree therefore store the encryption keys required to decrypt their children. The key that decrypts the root node of the B-Tree, however, is never stored on the persistent medium; the root key is only stored on the securely deleting medium. Only one such key is stored at any time. Old keys are securely deleted and replaced with a new key.

In addition to encryption, each node also stores the *cryptographic digest* (henceforth called *hash*) of its children for integrity in a straightforward application of a Merkle tree [103]. An authentic parent node guarantees the authenticity of its children. The root hash is stored with the key.

We use a form of *shadowed updates* [100] when updating B-Tree nodes. A shadowed update means that when a new version of a node is written, it is written to a new location. A node that references it (i.e., its parent) must also be updated to store the new location, propagating shadowing to the root. We use a variation on shadowed updates: instead of updating the location of the data, we instead update the key that decrypts it. Consequently, any change to a leaf node results in new versions of all ancestor nodes up to the root. This is analogous to normal shadowed updates if one imagines the encryption key as a pointer to the data's location.

10.4.1 Cryptographic Details

All encrypted data—both the B-Tree's node data and the user's actual data—are encrypted with AES keys in counter mode with a static IV. Keys are randomly generated using a cryptographically suitable entropy source. We use each key only once to encrypt data. Therefore, an encryption key's lifetime is the following: it is generated randomly, it is used once to encrypt data, and then it is used arbitrarily many times to decrypt that data until it is securely deleted.

10.4.2 Data Integrity

To ensure that the persistent storage has correctly returned all data, it is necessary to verify the integrity of the data received. Merkle Hash Trees [103] provide such a construction for binary tree data structures: each node in the tree contains a hash of the concatenated values of its two children, and an authentic copy of the hash of the root is sufficient to verify the integrity of all nodes in the tree. Mykletun et al. propose extending this to provide integrity for the many children of a B-Tree [104]. The use of a cryptographic hash function ensures that the received data is protected against both accidental and deliberate modification.

We use a variant of this approach in our solution: each node is hashed but the hashes of the children are independently stored in their parent (alongside the child's decryption key). Therefore, the parent stores for each child a key, a hash, a search index, and a storage location. The leaves of our B-Tree then store the hashes of the actual data blocks they index. Hashes for data blocks are computed when they are written, and all other B-Tree nodes have their hashes recomputed before writing them to persistent storage. This allows efficient updates to B-Trees with large numbers of children because only the children that actually changed need to have their hashes recomputed, not all children of a node. Moreover, it is these modified nodes that are available in the user's skeleton tree and must be also re-encrypted during committing.

10.4.3 Versioning

Some cloud storage research focuses on ensuring that the most recent version of data is always returned to protect against remote storage media that may return previous versions of valid data instead of the most recent. For instance, tahoefs [105] queries a variety of servers and each returns a data block and a version; tahoefs assumes that the highest received version number is the correct one.

A nice property of our solution is that whenever a data block is updated, the old version becomes irrecoverable even to the user. This means that our solution achieves authenticated versions as a side effect. If a data block can be correctly decrypted, it is therefore the newest version.

10.4.4 Skeleton Tree

All the B-Tree nodes are stored on the persistent storage. To improve efficiency, however, the actual B-Tree operations are performed on a smaller subset of the B-Tree cached in memory, which is called the *skeleton tree*. The skeleton tree reduces the cost of computing decryption keys for data when the relevant B-Tree nodes are available in memory; this strongly benefits, in particular, sequential data access. It also permits multiple updates to the B-Tree to be batched and committed together, which reduces the total number of B-Tree nodes to rekey. Finally, it allows the user to control how often the securely deleting storage medium is updated (i.e., the clock period and relevant latencies). This is useful if using the medium's deletion operation has a non-trivial cost in latency, wear, or human effort.

Initially, the skeleton tree only stores the root of the B-Tree; other node references are loaded lazily. Figure 10.1 gives an example of this configuration, where the persistent storage has a stale B-Tree and the skeleton tree reflects some combination of addition, removal, and rebalance operations. When a B-Tree operation requires accessing a node missing from the skeleton tree, the corresponding B-Tree node is read from persistent storage and decrypted. Its integrity is confirmed by using its hash value stored at the parent and the missing reference is added to the skeleton tree. This new reference now stores the decryption keys and integrity hashes corresponding to all its (missing) children, allowing the skeleton tree to grow further on request. The size of the skeleton tree is limited: when it reaches its capacity then nodes are evicted from the tree. In Section 10.6 we present our experimental results with eviction strategies.

All B-Tree modifications—e.g., deleting data and rebalancing—are performed on the skeleton tree and periodically committed in batch to persistent storage. A *dirty* marker is kept with the skeleton nodes to indicate which of them have local changes that need committing. Whenever a tree node is mutated—i.e., adding, removing, or modifying a child reference—it is marked as dirty. This includes modifications made due to rebalance operations. B-Tree nodes that are created or deleted—

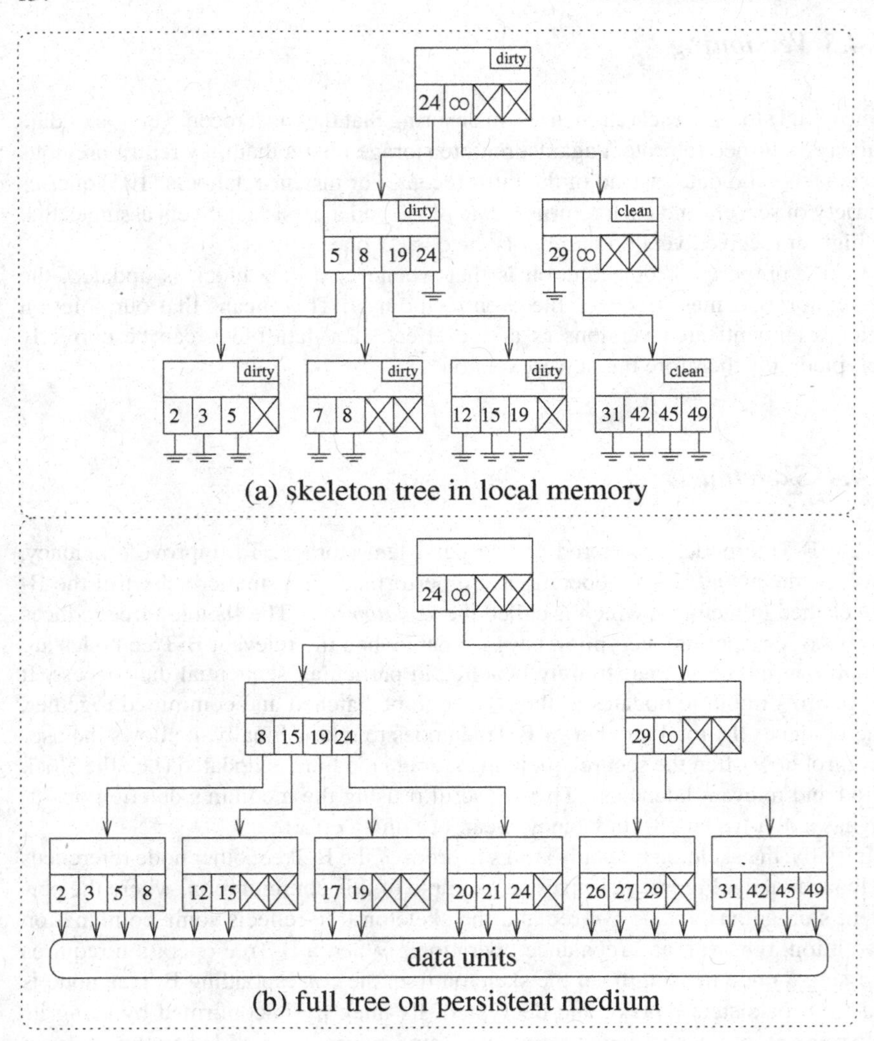

Fig. 10.1 Example of a B-Tree stored on the persistent medium along with an in-memory skeleton tree. (a) shows the skeleton tree of B-Tree nodes, where node 42 was read and local changes were made: the node 7 was added and the node 17 was deleted, causing a split operation and a fuse operation, respectively. (b) shows the persistent medium which stores all the nodes in the tree, some of which are stale. Only the nodes that have been needed are loaded into the skeleton tree.

due to splitting or fusing nodes—are also marked as dirty. Finally, dirtiness is propagated up the skeleton tree to the current root.

10.4.5 Commitment

The B-Tree commit operation is periodically performed and is the clock of our system. Time is thus divided into deletion epochs characterized by the particular master encryption key stored in the securely deleting storage medium. Commit writes new versions of all the dirty nodes to persistent storage, thus achieving secure deletion of deleted and overwritten data. Modifications to the B-Tree are first cached and aggregated in the skeleton tree, and then they are simultaneously committed. This means that deleted data items have a deletion latency bounded by the clock period, i.e., a SECDEL-CLOCK implementation.

The commit operation handles two kinds of dirty nodes: *deleted* ones that have been deleted from the B-Tree through the fuse operation, and *valid* ones that are still part of the tree. Each valid dirty node is first associated with a fresh randomly generated encryption key. Because parent nodes store the keys of their children, all parents of dirty valid nodes are updated to store the new keys associated with each child. After this, the sub-tree of valid dirty nodes is traversed in post order to compute each dirty valid node's integrity hash, which is then stored in the parent. The root node's key and integrity hash are stored outside the tree local to the user. The data for each valid dirty node (i.e., its children's keys, hashes, and search values) is then encrypted with its newly generated key and stored on persistent storage.

10.4.6 Crash Safety

A critical feature in the design of storage systems is crash safety, which aims to minimize the data loss due to unexpected events such as a system crash. Unsaved data may reside in memory buffers waiting to be committed; such data is lost in the event of power loss. Thus, the commit period is a trade-off between data loss risk and increased overhead. The overhead is induced by the cost of rekeying and storing dirty tree nodes. Therefore, we use a journalling mechanism that allows the recovery of uncommitted data. This removes the risk of data loss from the trade-off.

When data is written, its index reference and encryption key are written to a journal on the local storage medium. This permits the encrypted data stored on remote storage to be decrypted without updating the encrypted parents. Similarly, when data is deleted from the remote storage, a record of this deletion is made in the journal. The journal is securely deleted after flushing the tree and replayed whenever the server application is started. This results in the correct reconstruction of the B-Tree's internal state at the time of system crash.

If the securely deleting storage medium is too resource constrained to maintain an adequate journal, then the user can safely store all changes by including the data item's key directly wrapped by the current root key. This permits secure deletion because the user is assured that the old root key is destroyed at the next commit operation—the wrapped key is useful only in the event of power loss before that commit operation occurs. The consequence of this is that it introduces an existential latency for the data: the compromise of the root key before the data is written as well as continuous access to the persistent storage provides access to the data. In this case, the existential latency is bounded by the clock period and the B-Tree has the behaviour of a SECDEL-CLOCK-EXIST implementation.

Observe that while this direct wrapping is not a shadowing mutation, it is easy to show that applying it to a mangrove still preserves the key disclosure graph's mangroveness: only one vertex and one edge are added such that the new vertex has outdegree zero (preventing cycles) and indegree one; the indegree of all other vertices is unchanged.

10.5 Implementation Details

We have implemented our B-Tree-based solution. We use Linux's network block device (`nbd`), which allows a listening TCP socket to receive and reply to block device I/O requests. In our case, we have our implementation listening on that TCP socket. The `nbd-client` program and `nbd` kernel module—required to connect a device to our implementation and format/mount the resulting device—remain unchanged, ensuring that no modifications to the operating system are required to use our solution. Our implementation includes the encrypted B-Tree described in this chapter and interacts with a variety of user-configurable storage backends. Our implementation is written in C++11 and is freely available with a GPL version 2 license.

10.5.1 Data Storage

Our solution assumes the user's storage system divides the data into data items indexed by a handle for storage. There are different ways this can be implemented. A basic a key-value storage system stores data items as values, and keys (handles) reference this data. Data items can be entire files, components of files, etc. This allows our solution to implement an object storage device (OSD) [58]. If each data item is uniquely indexed by the B-Tree, then modifying the data item requires re-encrypting it entirely with a new key and updating its reference in the B-Tree. This inhibits the ability to efficiently securely delete data from large files such as databases.

Alternatively, data can be divided into fixed-size blocks indexed by the B-Tree. This facilitates random updates as only a fixed-size block must be updated to make any change to data. This is the construction we use in our implementation: a virtual array of data is indexed by offsets of fixed-size blocks and exposed as a block-device interface. This block-device can then be formatted as any block-based file system, which is then overlaid on the B-Tree. Sparse areas of the file system then do not appear as keys in the B-Tree; if written to, the corresponding keys are added to the B-Tree.

10.5.2 Network Block Device

The network block device is a block device offered by Linux. It behaves as a normal block device that can be formatted with any block-based file system and mounted for use. However, it is actually a virtual block device that forwards all block operations over TCP (i.e., reading and writing blocks, as well as trim and flush commands). The listening user-space program is responsible for actually implementing the block device.

By default, the `nbd-server` program uses a local file to implement the block device. This is similar to the loopback device (e.g., `/dev/loop0`), except that the `nbd-server` can run on a separate machine than the file that corresponds to the block device. The `nbd-client` program tells the running kernel how to connect to a `nbd-server` program and which `nbd` device number it should connect to it. After successfully executing `nbd-client`, the corresponding block device forwards all requests to the desired `nbd-server` program. This permits our solution to be easily integrated into any Linux system without modifying the operating system. By default, however, running `nbd-client` and mounting the device driver `nbd` as a file system requires root privileges. Our solution replaces only the user-level `nbd-server` tool with our B-Tree implementation.

10.5.3 Virtual Storage Device

While the default `nbd-server` program simply serves a local file as a block device, we wrote our own implementation of a virtual block device that interfaces with a variety of back-end storage media. The reading and writing of blocks pass through our shadowing B-Tree implementation. It uses block addresses as indices in the B-Tree; the data's remote storage location in that block address is kept in the leaves of the B-Tree. The user selects how the resulting data is stored, including data blocks for nodes and data (persistent medium) as well as the master key and integrity hash (securely deleting medium).

10.5.4 Caches

Our solution caches data in multiple locations. Two important caches are the skeleton B-Tree and a working space for the `nbd` device. The first ensures that the B-Tree's dirty nodes do not need to be flushed—and the root key changed—whenever data is stored or removed. The second ensures that if the `nbd` device sequentially issues many small read or write requests on the same stored block, then the block is only retrieved once.

In Section 10.6 we test a variety of cache sizes and eviction strategies to quantify the success of caching. In our final design, we use a simple least-recently used eviction strategy for our B-Tree cache. Our approach is modified from simple cache eviction because we require a skeleton tree and a full commit of all dirty nodes during the clock operation. We therefore only perform a full flush of all dirty cached B-Tree nodes whenever there is insufficient cache space to accommodate the requirements of the worst-case B-Tree operation. The clean nodes can then be heuristically evicted as needed when they are leaves in the skeleton tree.

10.6 Experimental Evaluation

In this section, we evaluate the performance of the B-Tree under different workloads and investigate how the performance can be improved through different caching strategies.

10.6.1 Workloads

We test our implementation's performance on a variety of different file system workloads. We used the `filebench` utility [101] to generate three workloads and we also created our own workload by replaying our research group's version control history. We used filebench's `directio` mode to ensure that all reads and writes are sent directly to the block device and not served by any file system page cache; similarly, we synchronized the file system and flushed all file system buffers after each version update in our version control workload. The workloads we use are summarized as follows:

- `sequential` writes a 25 GiB file and then reads it contiguously. This tests the behaviour when copying very large files to and from storage.
- `random_1KiB` performs random 1 KiB reads and writes on a pre-written 25 GiB file. This tests the performance for a near-worst-case scenario: reads and writes without any temporal or spatial locality.
- `random_1MiB` performs random 1 MiB reads and writes on a pre-written 2 GiB file. This tests the performance for random access patterns with a larger block size that provides some spatial locality in accessed data.
- `svn` replays 25 GiB of our research group's version control history by iteratively checking out each version. This test provides an example of a realistic usage scenario for data being stored on a shared persistent storage medium.

We run our implementation behind an `nbd` virtual block device, formatted with the `ext2` file system. We mount the file system with the `discard` option to ensure that the file system identifies deleted blocks through TRIM commands.

10.6.2 Caching

We experimentally determine the effect of the skeleton tree's cache size and eviction strategy. Using the sequence of block requests characteristic of each workload, we use our B-Tree implementation to output a sequence of B-Tree node requests. A B-Tree node request occurs whenever the skeleton tree visits a node; missing nodes must be fetched from the persistent medium and correspond to cache misses. Observe that for the same workload, the sequence of node requests will vary depending on the B-Tree's block size. We output one B-Tree node request sequence for each block size that we test. With this sequence of node requests, we then simulate various cache sizes and caching behaviours.

We test three different strategies: Bélády's optimal "clairvoyant" strategy [102], least recently used (LRU), and least frequently used (LFU). Bélády's strategy is included as an objective reference point when comparing caching strategies. We only maintain cache usage statistics for items currently in the cache.

Table 10.1 Caching hit ratio (%) for B-Tree nodes

		cache size: 10 items			**cache size: 50 items**		
workload	**block size**	**LRU**	**LFU**	**Bélády**	**LRU**	**LFU**	**Bélády**
seq	1 KiB	97.5	28.2	97.7	98.8	38.7	98.8
seq	16 KiB	99.7	47.7	99.8	99.9	99.0	99.9
rand (1 KiB)	1 KiB	11.8	15.3	19.8	26.4	26.3	33.5
rand (1 KiB)	16 KiB	49.3	40.1	58.2	63.4	67.2	70.3
rand (1 MiB)	1 KiB	97.5	25.4	97.7	97.9	38.0	98.0
rand (1 MiB)	16 KiB	98.6	62.6	98.7	98.9	98.3	99.1
svn	1 KiB	96.0	47.2	96.1	97.1	75.9	97.4
svn	16 KiB	97.8	81.2	97.8	98.6	96.2	99.0

The results of our experiment are shown in Table 10.1. We observe that caching nodes is generally quite successful; many of the workloads and configurations have very high hit ratios. This is because contiguous ranges of block addresses tend to share paths in the B-Tree. Consequently, the cache size itself is not so important; if it is sufficiently large to hold a complete path, then sequential access occurs rapidly.

LRU is generally preferable to LFU. The only exception is very small random writes with a small block size. This is because such writes have no temporal locality and so the frequency-based metric better captures which nodes contain useful data. For random-access patterns, the cache size is far more important than the eviction strategy, a feature also observable from Bélády's optimal strategy. For any form of sequential access, LRU outperforms LFU because LFU unfairly evicts newly cached nodes, which may currently have few visits but are visited frequently after their first caching. We see that LRU often approaches Bélády's optimal strategy, implying that more-complicated strategies offer limited potential for improvement.

10.6.3 B-Tree Properties

We investigate our system's overhead with regards to the fetching and storing of nodes that index the data. We now characterize this with regards to different workloads and parameters, expressing the results with the following metrics:

- Cache hits: percentage of B-Tree node visits that do not require fetching.
- Storage overhead: ratio of node storage size to data storage expressed as a percentage.
- Communication overhead: ratio of the persistent medium's communication for fetching and storing nodes compared to the sum of useful data read and written, expressed as a percentage.
- Block-size overhead: ratio of additional network traffic (beyond the I/O) for fetching and storing data compared to the sum of data read and written by the file system, expressed as a percentage. (This is based only on the block size and workload; it is independent of using a B-Tree.)

Additionally, we characterize the following B-Tree properties common to all workloads:

- Total data blocks: 25 GiB divided by the block size.
- Tree height: the height of the B-Tree that indexes the number of data blocks.
- Cache size (nodes): the fixed cache size of 8 MiB expressed as nodes that fit into that capacity.
- MiBs sharing path: the size of contiguous data whose blocks all share a unique path, that is, how much data is indexed by a single leaf node.

Table 10.2 shows the results of our experiments. We see that in all cases the storage overhead of the B-Tree nodes is a few percent and decreases with the block size. In all workloads except `random_1KiB`, the communication overhead is also reasonable. Large block sizes benefit the most from sequential access patterns, because a large block size means more sequential data can be accessed without fetching new nodes (e.g., using a block size of 256 KiB results in half a GiB of data indexed by the same path in the B-Tree). Degenerate performance is observed for our worst-case workload: where data blocks are accessed in a completely random fashion without any spatial or temporal locality. As expected, the block size overhead resulting from fetching unnecessary data shows a large amount of waste.

10.7 Conclusions

We designed, implemented, and analyzed a securely deleting B-Tree, whose design is taken from the space of securely deleting data structures that we prove secure in Chapter 9. It uses a local skeleton tree to cache all modifications and performs a period commit operation to synchronize them and securely delete discarded data. Our

Table 10.2 B-Tree secure deletion overhead

		B-Tree block size			
		4 KiB	**16 KiB**	**64 KiB**	**256 KiB**
general	total data blocks	6553600	1638400	409600	102400
	tree height	5	3	2	2
	cache size (nodes)	2048	512	128	32
	MiBs sharing path	0.16	2.65	42.6	682.5
sequent.	cache hits (%)	99.3	99.7	99.9	1
	storage overhead (%)	2.4	0.6	0.1	0.03
	comm overhead (%)	2.4	0.6	0.1	0.03
	block-size overhead (%)	0	5.3	26.3	58.1
rand 1k	cache hits (%)	64.7	59	43.2	73.8
	storage overhead (%)	2.4	0.6	0.1	0.03
	comm overhead (%)	1308.5	3129	8623.5	20671.4
	block-size overhead (%)	497.9	2293.2	9473	38191.8
rand 1m	cache hits (%)	99.2	98.9	96.5	95.5
	storage overhead (%)	2.47	0.59	0.14	0.03
	comm overhead (%)	4.9	3.7	7.8	17.7
	block-size overhead (%)	1	7.7	34.6	82.1
svn	cache hits (%)	99.2	98.9	96.5	95.5
	storage overhead (%)	1.74	0.42	0.1	0.02
	comm overhead (%)	4.4	4.9	5.4	2.6
	block-size overhead (%)	0	63.4	247.9	750.2

analysis showed that the communication and storage overhead is typically negligible and the skeleton tree's caching of B-Tree nodes is very effective.

In our B-Tree, a single master key is stored on the securely deleting storage medium and is required to access all stored data. In the next chapter, we consider the problem of an unreliable securely deleting storage medium, i.e., one that may fail to be available, to correctly store data, or correctly delete data. We design a robust storage medium to account for these risks.

10.8 Practitioner's Notes

- Our network block device implementation allowed the secure deletion to work independent of files. Access patterns for different file systems, e.g., due to the storage location for metadata, may not be optimal with regards to the actual secure-deletion structure and this should be optimized.
- In our experiments, the workload closest to cloud-based storage is the replay of our version control repository. Organizations with access to real-world workloads should experiment with those when tuning parameters.

- It is best practice to use the largest key size available in case of a novel cryptanalytic attack that weakens the security of a key.
- Note that authenticated encryption may be used to ensure the integrity of the B-Tree by incurring ciphertext expansion instead of the Merkle-like hash construction used. This also achieves the most recent version provided that each encryption key is only used to encrypt one piece of data.

Chapter 11
Robust Key Management for Secure Data Deletion

11.1 Introduction

As we have seen throughout the previous chapters, data encryption is a useful tool for reducing the problem of secure data deletion to the problem of deleting the corresponding encryption keys. These keys are smaller and so more easily managed and controlled than the data itself; they are stored only on storage media that provide secure deletion. In related work [2, 5, 10, 92], as well as our own work from Chapters 6, 7, 9, and 10, the securely deleting storage medium is assumed to have perfect storage characteristics: it never loses data, never exposes data except through compromise, it always correctly deletes data, and it is always available. These are strong and unrealistic assumptions to place on a storage medium. Moreover, the risk of data loss is amplified by the ratio between the size of the key and the data it encrypts. In particular, Chapter 10's B-Tree-based design used only a single encryption key to encrypt all the data stored on the persistent storage. The loss of these 16 bytes is devastating to such a system.

In this chapter, we remove the strong and unrealistic assumptions on the securely deleting medium's reliability, integrity, and confidentiality. We allow the securely deleting medium to be unavailable or to partially fail—either by discarding valid data or by failing to securely delete discarded data. We explore the effect that this has on the secure deletability and availability guarantees on the securely deleting storage medium, and we propose a system robust against such failures.

The system we use is a DNEFS-like keystore (Chapter 6) that provides securely deletable key values (KVs). These KVs are used to create encryption keys that encrypt data stored on the persistent storage, but the KVs are only ever stored by the keystore. Now we improve the robust storage and secure deletion of data by distributing the keystore over multiple nodes. This is particularly challenging since storing additional copies of data decreases the chance of its loss but increases the chance that one copy's deletion fails. At a minimum, we require that the failure or compromise of any single node in the system has no effect on the secure deletability or availability of data. We present a two-dimensional secret-sharing and erasure

J. Reardon, *Secure Data Deletion*, Information Security and Cryptography,
DOI 10.1007/978-3-319-28778-2_11

code that balances these properties. In one dimension, key values are replicated across a subset of nodes; in the other dimension, multiple key values are used to create encryption keys. We ensure that multiple key values can be selected such that no single node in the network stores any two of them.

We implement both our distributed keystore system and a FUSE-based [106] file system that uses the keystore to store securely deletable data with a variable encryption granularity. We test our design and measure our system's performance. We find that the latency of key operations is small and that the service rate is high; the latency remains small even as the service rate reaches its capacity. Individual nodes can service 14,000 requests per second with a latency (including local network delay) of approximately 725 μs. Moreover, the communication among keystore nodes is small, requiring, for example, 4 KiB/s to generate and synchronize millions of keys every 10 minutes.

11.2 System and Adversarial Model

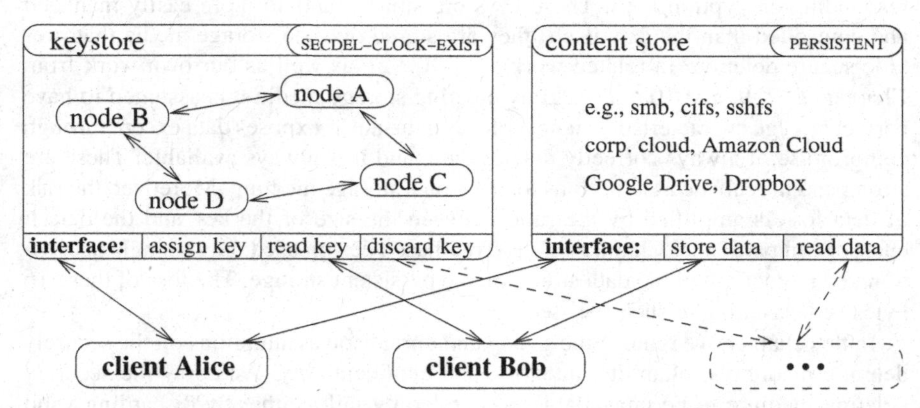

Fig. 11.1 Diagram of the keystore system. Each arrow is a mutually authentic, forward-secure communications channel. Multiple clients interact both with a content store and a distributed keystore. The content store has a PERSISTENT implementation. The keystore consists of a set of keystore nodes which mutually communicate and together implement a SECDEL-CLOCK-EXIST.

Our system model is similar to the one described in Chapter 9. There exist both a securely deleting storage medium and a persistent storage medium. In contrast to Chapter 9's model, however, the securely deleting storage medium is distributed among a set of *keystore nodes* that together implement a *clocked keystore*. Additionally, there is now more than one user of the system; we use the term *clients* to refer to them. The persistent storage medium is shared among the clients and is called the *content store*.

Figure 11.1 illustrates our system model. Each arrow is a mutually authentic, forward-secure communication channel, that is, a secure channel offering secrecy and authentication that also achieves perfect forward secrecy. This section overviews each entity and presents our adversarial model. Further details on components are presented in the subsequent sections: Section 11.3 describes the keystore, Section 11.4 describes how keystore nodes communicate, and Section 11.6 describes how the client uses these components to store data in a securely deletable way.

11.2.1 System Entities

Keystore.

Chapter 6 introduced the notion of a *clocked keystore*, which we revisit in this chapter. A keystore provides securely deletable key values (KVs), which can be used as encryption keys. Discarded KVs are periodically securely deleted, which is the *clock* aspect of the clocked keystore.

In this chapter, the *keystore* now comprises a set of *keystore nodes*, which together implement the clocked keystore. We assume that keystore nodes can authentically and forward-securely communicate between themselves. Authenticity means that communicating nodes know their partner's identity and agree on data sent. Forward security means that the compromise of a node's long-term key does not reveal the content of previous communications secured using that key. For example, TLS [107] with both client authentication and Diffie-Hellman key negotiation [108] satisfies these properties, provided that each node can verify the other nodes' certificates. We further assume that all nodes can store data such that it is securely deletable, i.e., each node has access to a securely deleting storage medium. We assume that this securely deleting storage medium may occasionally fail to correctly store data, delete data, or keep data confidential.

Content Store.

The *content store* is the persistent storage medium that stores encrypted data. This can be implemented as, for example, a virtual network file system. The content store also performs appropriate access control on the data: authenticating clients and checking permissions to access stored items. The content store is assumed to store data persistently in that stored data is never deleted. While truly persistent media are unrealistic, this is meant to model worst-case secure data deletion in practice—particularly when the content store is adversarially controlled. Its persistence is not required to work, but is rather a worst-case scenario.

The functionality of the content store remains unchanged. The way that it handles data's erasure coding, striping, migrating, defragmenting, backing up, etc., is unaffected by our solution. For example, the content store may bundle unrelated backups into replication sets such that a missing element may be reconstructed from the remaining elements; *deleting* a backup therefore requires reconstructing the replication set with a new element. By removing any requirement for secure deletion on the content store, encrypted data corresponding to *deleted* backups may remain on the content store and are replaced in replication sets when convenient.

Client.

The *client* stores and retrieves data. There can be multiple clients using the system, and clients must not be able to access other clients' data. We assume that clients can establish an authentic and forward-secure connection to the *keystore nodes* and the *content store* [107]. This can be scalably achieved in various ways. Our implementation uses a public-key infrastructure with a certificate authority [108, 109]; clients then verify the certificates of the entities to ensure that the public keys are valid before negotiating a TLS-secured communication channel.

11.2.2 Adversarial Model

We have the same concept of data items, data lifetimes, and secure deletion as the other work in this book. We assume the existence of a computationally bounded unpredictable multiple-access coercive adversary who can gain access to the keystore nodes and clients' secrets that may be needed to access their data.

Our adversarial model augments the adversary with the following *non-coercive attacks*:

- **A1** The content store is perpetually compromised. Data stored on it is immediately given to the adversary. This models settings where the content store is implemented using an untrusted or potentially compromised third-party storage system, such as Amazon Cloud [110], Dropbox [111], or Google Drive [112].
- **A2** There are multiple legitimate clients who have access to some, but not all, data. The adversary may be a legitimate client (or a coalition thereof) and thus use client privileges to access other clients' data.
- **A3** Some keystore nodes may fail. In particular, they may fail to securely delete stored data, fail to correctly store data, fail to maintain the confidentiality of data, fail in their availability, e.g., by being unable to respond to requests to delete data, and fail by generating predictable random numbers.
- **A4** There is a computationally bounded active network adversary for all communication: client to content store, client to keystore node, and between pairs of keystore nodes. We assume that a long-term denial-of-service attack on all keystore nodes is outside the adversary's abilities.

Because these attacks are non-coercive, the adversary can perform them at any time without it being considered a compromise; this is to say that data considered valid at the time these attacks occur is still eligible to be securely deleted.

11.3 Distributed Keystore

In Chapter 6 we introduced the notion of a clocked keystore, which provides individually assignable securely deletable *key values* (KVs). KVs are used to generate encryption keys. Each key has a corresponding *access token* (AT). Each data item is encrypted with its own unique KV, and secure deletion is achieved against a computationally bounded adversary by securely deleting the KV that corresponds to the data item.

The following properties ensure that this system provides secure data deletion at a fine granularity:

- **P1** The KVs associated to the ATs returned by `assign` must be unpredictable with a negligible guessing probability decreasing exponentially in κ.
- **P2** KVs returned by `read` must be the same for a particular AT from the time `assign` returns it until it is provided to `discard`; further, `read` must not return $\perp$ during this time.
- **P3** KVs returned by `read` must be unpredictable given the keystore's state at all times before the time `assign` returns its corresponding AT minus a bounded existential latency, and at all times after the time `discard` is called with its corresponding AT plus a bounded deletion latency.

In this section, we design a *distributed clocked keystore* that is robust against partial failures in the storage medium. It is composed of a set of keystore nodes. Each node maintains a local state, which includes the state and value for a set of key positions (KPs). We describe the keystore nodes' client interface, and how they together effect a clocked keystore. We then present the distributed synchronization method, and measures to detect and correct for Byzantine failures.

In our setting, multiple clients share access to the keystore. It is therefore necessary to provide access control on KVs. We use unpredictable access tokens to accomplish that. As such, we require an additional keystore property:

- **P4** ATs returned by `assign` must be unpredictable.

This provides the condition that knowledge of an AT allows knowledge of the corresponding KV.

11.3.1 Distributed Clocked Keystore

We now explain how to distribute our clocked keystore implementation over multiple keystore nodes and achieve correctness and robustness. Each keystore node maintains a local state, which includes the state and value for a set of positions. Table 11.1 lists the node's local state used by our algorithms. In it, Π represents the space of possible KPs, κ is a security parameter, and δ is the desired bound for existential and deletion latencies. Recall from Chapter 6 that the possible states for a KP are unused (**U**), assigned (**A**), and discarded (**D**).

As before, each node offers the client a keystore interface: assign a KV, read the KV associated with a KP, and discard the KV associated with a KP. Nodes are, however, now responsible for only a subset of KPs; the set of nodes responsible for a KP is called the KP's *replication set*. Reading a KP for which a node is not responsible returns $\perp$. Algorithms 1, 2, and 3 provide the client-side assign, read, and discard algorithms, respectively. The DOM function takes the domain of a mapping. The **with** operator indicates arbitrary selection. The **now** operator provides the current time.

Table 11.1 Keystore node local state.

Π	// set of KPs
κ	// security parameter
δ	// bound for latencies
St {	
me: $\mathbb{Z}^n$	// keystore node number
PK: $\mathbb{Z}^n \to$ pubkey	// public keys for peers
assigner: $\Pi \nrightarrow \mathbb{Z}^n$	// maps position to unique assigner
replicators: $\Pi \nrightarrow 2^{\mathbb{Z}^n}$	// maps position to set of replicators
state: $\Pi \nrightarrow \{\mathbf{U}, \mathbf{A}, \mathbf{D}\}$	// maps position to state
value: $\Pi \nrightarrow \{0,1\}^\kappa$	// maps position to KV
update_number: $\Pi \nrightarrow \mathbb{Z}^+$	// maps position to update count
update_commit: $\Pi \nrightarrow \{0,1\}^\kappa$	// maps position to current commit
last_update: $\Pi \nrightarrow \mathbb{Z}^+$	// timestamp of last update
updating: $\Pi \nrightarrow \{\textbf{true}, \textbf{false}\}$	// is position being updated
// check that the position stores a fresh value	
is_recent(t) $\triangleq$ ($\textbf{now} - t < \delta$)	
// the set of positions that are assignable for this node	
assignable $\triangleq$ $\{i \in \Pi : \text{state}(\pi) = \mathbf{U} \wedge \text{assigner}(\pi) = \text{me}$ $\wedge\ \text{is\_recent}(\text{last\_update}(\pi)) \wedge \neg\text{updating}(\pi)\}$	
}	

11.3.2 Distributed Keystore Correctness

In order for our distributed system to implement a keystore, it must guarantee properties **P1–3** as well as **P4** for our shared setting. An immediate problem with assignment of KPs in a distributed system is ensuring that two nodes do not assign the same KP. We remedy this by associating each KP with a unique *assigner*. Only the

Algorithm 1: assign

local state: St – keystore node state (Table 11.1)
output : AT or $\bot$ – access token or fail
begin
  if *St.assignable* $= \emptyset$ **then**
    return $\bot$;
  with $\pi \in$ *St.assignable* **do**
    St.state(π) $\leftarrow$ **A**;
    return $\pi \| \texttt{hash}(\text{St.value}(\pi))$;

Algorithm 2: read

local state: St – keystore node state (Table 11.1)
input : $\pi \| t$ – access token (position and hash)
output : KV or $\bot$ – key value for the position or fail
begin
  if $\pi \notin \text{DOM}(St.value)$ **then**
    return $\bot$;
  if $\texttt{hash}(St.value(\pi)) \neq t$ **then**
    return $\bot$;
  return *St.value*(π);

Algorithm 3: discard

local state: St – keystore node state (Table 11.1)
input : $\pi \| t$ – access token (position and hash)
output : $\top$ or $\bot$ – success or fail
begin
  if $\pi \notin \text{DOM}(St.value)$ **then**
    return $\bot$;
  if $\texttt{hash}(St.value(\pi)) \neq t$ **then**
    return $\bot$;
  St.state(π) $\leftarrow$ **D**;
  return $\top$;

assigner may assign the value and all other nodes that store the value are *replicators*; a KP's *replication set* therefore consists of a single assigner and some number of replicators. The set of assignable positions among keystore nodes are pairwise disjoint. This restriction does not limit our solution because the client has no preference over assignable positions: the client asks a node for an unused KV and accepts it. Nodes create their own assignable KPs as necessary and cooperate to replicate them, ensuring that all nodes can satisfy a client's assign request.

To see why **P1** holds, first observe that it holds for each node individually since **U** KPs store unpredictable KVs and are changed to **A** when assigned; the KV is replaced with a new value when the corresponding KP returns to **U**. Second, each keystore node is given a unique set of assignable KPs, so given all assignable KVs from all other nodes, the assignable KVs from the remaining node are still unpredictable. **P2** holds because nodes in the replication set can read the KV. **P3** holds because a distributed synchronization protocol (described in Section 11.4) is periodically executed among nodes that store the KP, which implements the clock function of the keystore. **P4** holds because ATs include the cryptographic hash of the corresponding KV—itself unpredictable through **P1**.

Note that **P1–3** are only provided assuming that no node fails. We now describe the synchronization protocol that runs among the nodes in the replication set and then consider the Byzantine failures that can occur, and design our system to be robust against them.

11.4 Synchronization

Keystore nodes use a *synchronization* protocol to ensure that KPs have consistent states and values. Synchronization results in the distributed keystore having the behaviour of a simple keystore and therefore achieving properties **P1–4**. Synchronization is an assigner-led two-phase protocol, similar to the two-phase commit atomic commitment protocol [113]. The first *pull* phase (cf. voting) collects all state changes from replicators. The second *push* phase (cf. commit) provides replicators with the pulled information, allowing them to all independently agree on the new state and value.

Algorithm 4 presents the synchronization procedure as run by the assigner. The **assert** keyword represents a check or condition that must hold; if any asserted statements fail—e.g., because a signature is forged—then the update is aborted. We note that while the algorithms presented here describe the synchronization of a single key position, our implementation improves the communication complexity by synchronizing many key positions simultaneously. We restate that communication between nodes occurs over mutually authentic forward-secure TLS-encrypted communication channels.

In the *pull* phase the assigner requests the current state from each replicator as well as a random contribution used to generate the new KV. Algorithm 5 presents the replicator-executed pull algorithm. The assigner commits to its random contribution at this time, which the replicators store to check during the push phase. Each replicator returns its identity, the synchronization round number, the KP's ID and state, a random contribution, and the assigner's commitment, as well as a public-key signature of this data. These pulled values are collected for all nodes in the replication set and distributed to each (including the assigner) during the subsequent *push* phase.

Algorithm 6 presents the push algorithm executed by all nodes in the replication set. It takes as arguments the KP to push, the assigner's previously committed random value, and the set of pull messages aggregated by the assigner. It checks that only nodes in the replication set appear in the set of messages and that each node appears exactly once, that each message concerns the correct position and synchronization round number, that the assigner's random contribution agrees with its previous commitment, that all nodes were given the same commitment, and that each push message is correctly signed by the providing node. If so, then the vote-state algorithm is used to update the state and determine if the value must be replaced, using the same procedure as the assigner. The push algorithm returns the hash of the update number, position number, and the value. The assigner checks that each replicator's hash matches its own version.

Algorithm 4: synchronize

```
local state: St – keystore node state (Table 11.1)
           : peer – map from node number to peer for communication
input      : π – key position to synchronize
begin
    assert St.assigner(π) = St.me;
    atomic
        if St.updating(π) = true then
            return
        St.updating(π) ← true;
    /* setup                                                    */
    St.update_number(π) += 1;
    i ← St.update_number(π);
    k_me ← random-key();
    c ← hash(k_me);
    s ← St.state(π);
    St.update_commit(π) ← c;
    /* pull phase                                               */
    R ← [(π, me, i, s, k, c, sign(π||me||i||s||k||c))];
    for r ∈ St.replicators(π) do
        R.append(peer(r).pull(i, π, c));
    /* push phase                                               */
    h ← peer(St.me).push(π, R);
    for r ∈ St.replicators(π) do
        h' ← peer(r).push(π, R);
        assert h' = h;
    St.updating(π) ← false;
```

The use of multiple random contributions from the nodes protects against a case where an assigner has a broken random number generator that produces predictable numbers. While making the assigners commit to their random key values is not strictly necessary in our security model, it ensures that the assigner—or indeed any single entity—cannot dictate new key values. In Section 11.6.4 we discuss a scenario where this is useful.

Algorithm 7 presents the vote-state algorithm used to determine the new key state and value. It takes a set of states from among the replication set and its own local state. It returns the new state and whether it should be replaced with a new

Algorithm 5: pull

```
local state: St – keystore node state (Table 11.1)
input    : i – update number
         : π – key position to pull
         : c – commitment for push phase
output   : π – key position being pulled
         : r – my keystore number
         : i – output number
         : s – position state
         : k – random contribution
         : σ – signature of (π||r||i||s||k)
begin
    r ← St.me;
    assert St.update_number(π) + 1 = i;
    St.update_number(π) ← i;
    St.update_commit(π) ← c;
    s ← St.state(π);
    k ← random-key();
    return (π, r, i, s, k, c, sign(π||r||i||s||k||c));
```

Algorithm 6: push

```
local state: St – keystore node state (Table 11.1)
input    : π – key position to push
         : R – list of pull messages
output   : h – hash of new key value
begin
    P ← St.replicators(π) ∪ St.assigner(π);
    k ← 0;
    S ← [];
    for (π′, r′, i′, s′, k′, c′, σ′) ∈ R do
        assert r′ ∈ P;
        P ← P \ {r′};
        assert c′ = St.update_commit(π);
        if r′ = St.assigner(π) then
            assert St.update_commit(π) = H(k′);
        assert π′ = π;
        assert St.update_number(π) = i′;
        assert verify(St.PK(r′), π′||r′||i′||s′||k′||c′, σ′);
        k ← k ⊕ k′;
        S.append(s′);
    assert P = ∅;
    (s_new, replace) ← vote-state(S);
    if replace = true then
        St.value(π) ← k;
        St.last_update(π) ← now;
    h ← hash(i||π||St.value(π));
    return h;
```

Algorithm 7: vote-state

```
local state: q_a – assign quorum
           : q_d – discard quorum
input      : {s_1,...,s_n} – replication set's states
           : s_me – my state
output     : s_new – new state of position
           : replace – whether the key value must be replaced
begin
    n_a ← |{i ∈ [1,n] : s_i = A}|;
    n_d ← |{i ∈ [1,n] : s_i = D}|;
    if n_d ≥ q_d then
        return (D, true);
    if n_a ≥ q_a then
        return (s_me = D ? D : A, false);
    return (D, true);
```

value. Further details are presented in the next section on Byzantine robustness. In principle, the algorithm works by requiring a quorum of nodes to agree on assigns and discards. Each node votes on whether a position is **D**; if the vote passes—that is, the *discard quorum* is met or exceeded—then the position is discarded and replaced. If not, then a vote occurs on whether the position is **A**; if the vote passes—that is, the *assign quorum* is met or exceeded—then the position is **A** and the value is retained. Otherwise, if neither vote passes, the position is considered **D** and replaced. Positions that stay unused until the next clock edge are considered discarded and they are replaced to ensure that the KVs have bounded existential latencies. Observe also that a successful discard vote overrules a successful assign vote.

Figure 11.2 shows an example state timeline for a KP stored by an assigner and two replicators. The nodes in the replication set periodically synchronize. The thick lines for the assigner and replicators represent a concurrent example history for the KP's state. The final sequence consists of the KP being assigned and deleted by the assigner before synchronizing.

11.5 Byzantine Robustness

Our distributed keystore provides properties **P1–4** under perfect conditions. In this section we make it robust against Byzantine failures. A Byzantine failure in a distributed system is a failure where a node behaves in an arbitrary way [114]. Table 11.2 lists the Byzantine failures during the synchronization protocol. The first part of the table lists all possible false state reports from nodes by enumerating them. For example, F1 is when a KP's state is **U** but a node reports that it is **A**. The effect column shows the effect of this failure, and the remedy column shows how it is avoided. For later clarity, *fraudulent failures* are where a node reports a state change

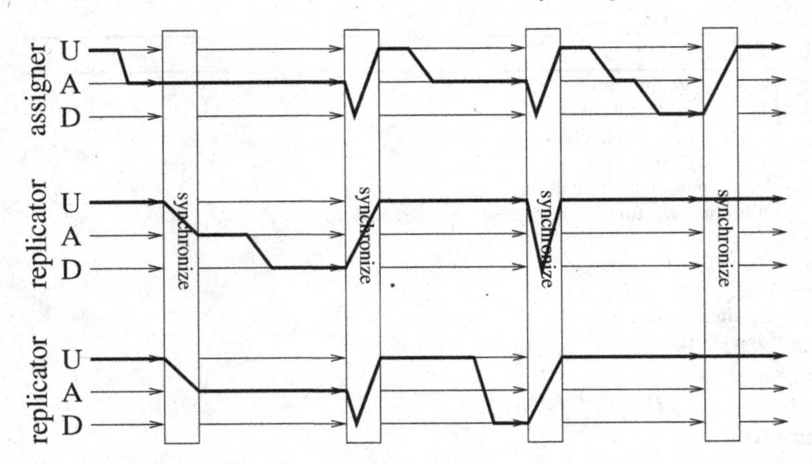

Fig. 11.2 Example position-state timeline for a three-node replication set. The position is assigned then discarded, thrice. This example uses an assign and discard quorum of one.

that *did not* occur, and *negligent failures* are where a node does not report a state change that *did* occur. The second part of the table lists non-state-related Byzantine failures found by inspecting the algorithms.

A node may also have a Byzantine failure when serving a client's requests, however most of these are equivalent to a particular failure during synchronization. For example, during assign, a node may return incorrect data to the client—this is equivalent to F9: the node gives a false random seed to other nodes during synchronization and thereby updates the KV to a different value. Table 11.3 presents the different Byzantine failures that can occur for client operations by inspecting Algorithms 1–3. It also shows the synchronization failure to which they are equivalent, and the remedy that makes the keystore robust against it.

Replication Set Eviction.

F7 and F8 are the failures where nodes delay or disrupt synchronization, such as by being offline or by sending bad signatures. An assigner may delay initiating or completing synchronization; replicators may delay responding to pull requests. Both delays can bring about a longer deletion latency, however existential latency is unaffected because the definition of an assignable KP excludes ones that were not recently synchronized.

Delayed synchronization is remedied with replication set eviction. After a configurable amount of time passes without synchronization, unresponsive nodes are evicted from the replication set. The assigner then replicates the KP with new replicators at the next opportunity. Discarded KPs known locally to the evicted replicator are lost, but this is equivalent to F6. The offline node may still retain deleted KVs, however, but we discuss how to resolve this later. Unresponsive assigners can also

Table 11.2 Byzantine failures in keystore node synchronization.

ID	state change true	state change claim	effect	remedy
F1	**U**	**A**	needless key storage	load balance assigner pages
F2	**U**	**D**	no negative effect	discard quorum
F3	**A**	**U**	lost assigned key	over-report assigns, check value
F4	**A**	**D**	lost assigned key	discard quorum
F5	**D**	**U**	no negative effect	discard quorum
F6	**D**	**A**	keep discarded key	over-report deletions

ID	Byzantine failure	effect	remedy
F7	delay synchronization	increase latencies	peer eviction
F8	send bad signatures	increase latencies	peer eviction
F9	false random seed	assigned keys differ	over-report assigns
F10	send false pull msg	fake consensus	peer signature
F11	replay pull msg	fake consensus	sign over update number & position
F12	(i) missing pull msg (ii) duplicated pull msg	fake consensus	check pull messages against replication set

Table 11.3 Byzantine failures in keystore node client interface.

operation	Byzantine failure	equiv.	remedy
read	deny service	-	multiple nodes
	return garbage	F9	client check
assign	deny service	-	multiple nodes
	unfresh value	-	discussed in Section 11.6.4
	multiple assign	-	discussed in Section 11.6.4
	do not update state	F3	over-report assigns
	return garbage	F9	over-report assigns
discard	deny service	-	multiple nodes
	do not update state	F6	over-report discards
	allow fake token	F4	deletion quorum

be evicted. This results in assignments being unreported (equivalent to F3) and discards being unreported (equivalent to F6).

Over-Reporting Thresholds.

Clients over-report their assigned and discarded KPs to multiple nodes in the replication set to protect against negligent failures. Two configurable *over-reporting thresholds*, r_a and r_d, are the number of additional nodes to which each assigned or discarded KP is reported, respectively. When over-reporting an assign, the AT further serves as the proof that the assigned KV is in fact the same one stored by the replicating node, thus simultaneously protecting against providing false data (F9).

Assign and Discard Quorums.

We are robust against fraudulent failures by setting a configurable *assign quorum* and *discard quorum*. These quorums, denoted q_a and q_d respectively, were introduced by the vote-state algorithm (Algorithm 7), which checks the assign and discard counts against the relevant quorums during synchronization. The assign quorum is the number of nodes that must agree that a KP was assigned to change its state during synchronization; the discard quorum is the number of nodes that must agree that a KP was discarded to securely delete it during synchronization.

We emphasize that $q_a > 1$ implies that KVs for assigned KPs can be lost during synchronization and so we use $q_a = 1$ as a safe default. If fraudulent assigns are a concern, however, then it is necessary that the client only uses the KV after confirming the assignment through over-reporting. Discard quorums are important to protect against data loss due to a fraudulent failure. Assign quorums, however, only protect against wasted storage resources for KVs that were not actually assigned.

Summary.

Quorums and over-reporting place the onus on the client to report assigns and discards to sufficiently many nodes to meet the threshold; clients report assigns to replicators to ensure that the KV is correct and report discards to ensure that the state change occurs.

An assign and discard quorum of q_a and q_d, respectively, means that the system can handle at most $q_a - 1$ fraudulent assigns and $q_d - 1$ fraudulent discards. An over-reporting threshold of r_a (for assigns) and r_d (for discards) means that the system can handle at most r_a and r_d negligence failures for assigns and discards, respectively. Clients communicate with $q_a + r_a$ nodes per assign and $q_d + r_d$ nodes per discard.

11.6 Keystore Secure Deletion

In the previous section we saw how to implement a distributed securely deleting keystore that provides securely deletable key values (KVs). We now present how to use such a keystore to achieve secure deletion. We discuss policies on KVs, encryption key construction, and present an example encoding scheme that balances secure deletion and availability of data. Finally, we analyze the system's security.

11.6.1 Key Pools and Encryption Keys

Clients use the keystore to obtain securely deletable KVs that are used to build encryption keys. Different properties may be associated with different KVs. For example, KVs can have different deletion latencies, expected reliabilities, and procedures for their storage and secure deletion. All policy-based aspects by which KVs may be discriminated is encapsulated in the KV's *key pool*, which is a named set of KVs. When a client requests a KV, it does so by specifying the corresponding key pool from which the KV should be taken. Unused KVs are therefore fungible within a key pool.

We say a keystore node *serves* a key pool if it stores KVs for that pool. A node can serve multiple pools, but to serve a pool it must fulfill the pool's policy requirements. This therefore requires oversight in the initial deployment of keystore hardware, e.g., where they are located and how they store and securely delete KVs. Keystore node certificates also bind their service to key pools. Communication between two nodes is only necessary if the nodes serve the same pool. Figure 11.3 illustrates a system of six nodes serving 24 KVs from four pools.

An important notion for our system is that of *complementary key pools*: two key pools, L and L', are *complementary* if no keystore node serves both L and L'. This provides clients with the technical means of selecting two KVs with the knowledge that no correct keystore node stores both these KVs. Building an encryption key from the logical XOR of these KVs therefore ensures that the compromise or failure of any single machine does not affect the secure-deletability of any encryption key.

A *key recipe* or *recipe* is a sequence of ATs. On the content store, the key recipe is stored alongside the data encrypted with the resulting key. Clients are free to select

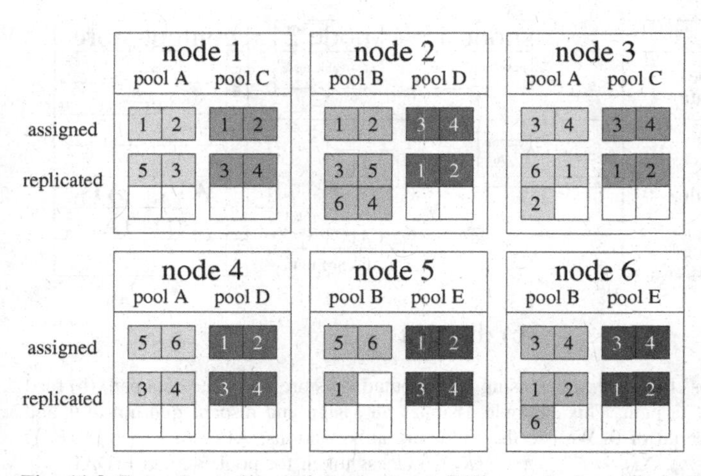

Fig. 11.3 Example pool-to-node assignment. Different shades indicate different pools, and the number corresponds to the pool-local key values. Each pool requires a minimum of one replication. No node serves both A and B; no node serves any two of C, D, and E. All nodes have two assignable KVs per pool and a number of replicated KVs.

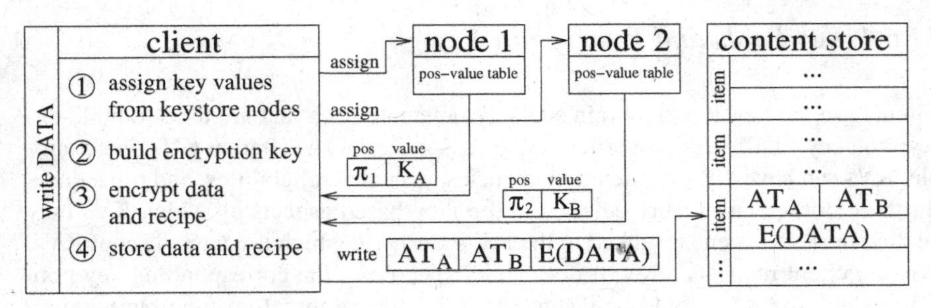

(a) write

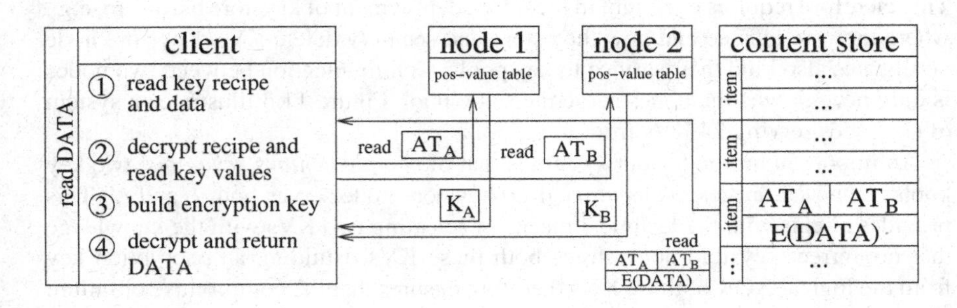

(b) read

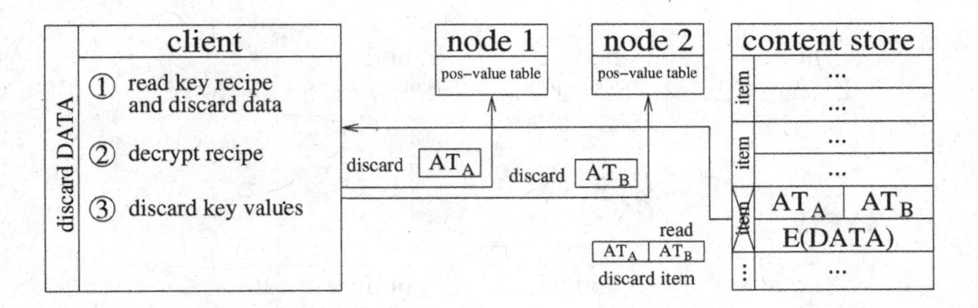

(c) discard

Fig. 11.4 Datapaths for client operations using a distributed keystore: (a) write data path (b) read data path (c) discard data path. This example assumes an assign and discard quorum of 1 and an over-reporting threshold of 0. We use the following abbreviations: $\mathrm{AT}_A = E_{K_{\mathrm{PK}}}(\pi_1 \| H(K_A))$ (access token for pool A), $\mathrm{AT}_B = E_{K_{\mathrm{PK}}}(\pi_2 \| H(K_B))$ (access token for pool B), and E(DATA) $= E_{K_A \oplus K_B}(\mathrm{DATA})$ (encrypted data).

how their encryption keys are composed, which they do by selecting the number of KPs and the pools from which to assign for each position. A sequence of pools that scaffolds the contents of a key recipe is called a *key class* or *class*.

The three client-side keystore operations—assign, read, and discard—extend naturally to classes and recipes. Clients assign a recipe from a class, read a key from a recipe, and discard a recipe. Figures 11.4 (a), (b), and (c) show the data path for the client write, read, and discard operations, respectively. When writing data, two KVs are assigned from keystore nodes, and the data is encrypted with a key derived from a key-encoding function applied to KVs. These KVs' ATs are then encrypted with a password-derived key [85] and stored alongside the encrypted data. The password ensures that the key recipes cannot be read. This is only important while the encrypted data is valid because compromised passwords cannot reveal securely deleted KVs.

When reading data, the key recipe provides the ATs required to obtain the KVs used to build the encryption key. When discarding data, only the recipe is read and is used to determine which keys to discard. Figure 11.4 is simplified in that the client does not over-report assigns and discards and only communicates with the KV's assigner.

In the next section, we discuss two different encodings that build actual encryption keys out of the keystore's KVs.

11.6.2 Encryption Key Encoding

Different encodings can be used to create cryptographic keys out of KVs. Encodings that require algebraic structure, such as polynomial-interpolation encoding [94], cannot be implemented using only client assigns: either all nodes involved must know the resulting encryption key to correctly generate shares or the client must generate additional shares and store them after the assignment with further nodes.

KVs are the atomic building blocks for encryption keys, and they can be combined in different ways. In this section we describe how to do this, presenting a simple XOR-based encoding scheme, a polynomial-interpolation encoding, and conclude with a comparison of the storage resources and failure robustness of these encodings.

11.6.3 XOR-Based Encoding

The XOR-based encoding for encryption-key generation consists of a client obtaining a set of KVs and deriving an encryption key from their logical XOR. This encoding has two parameters: the number of KVs XORed together and the number of times each KV is replicated. Both these parameters affect the availability and secure deletion of data. Figure 11.5 (a) illustrates these parameters as a two-dimensional

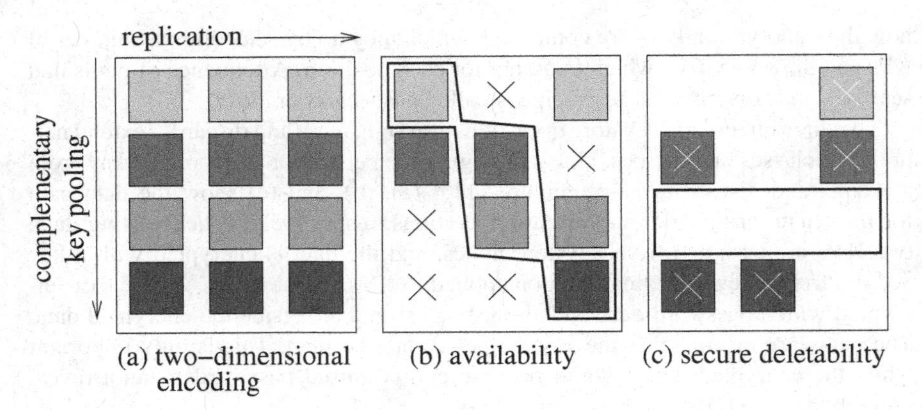

Fig. 11.5 Two-dimensional encryption key encoding using multiple KVs as XOR operands. Each square represents a single node in the system. Each row of nodes is a replication set for a particular KV. (a) Perfect replication for each KV. (b) Despite availability failures (denoted $\times$), each row's KV is still available and therefore so is the resulting encryption key. (c) Despite secure deletion failures ($\times$), the entire third row is securely deleted and therefore so is the resulting encryption key.

matrix. Each square represents a single node in the keystore; each row represents a replication set for a particular KV. There are four different KV stored among twelve nodes. This corresponds to an encoding where four XOR operands, each stored three times, are used to build encryption keys. Furthermore, complementary key pooling is used for each KV to ensure that a single node's failure affects only a single XOR operand, thus ensuring that encryption keys making use of it are not compromised.

For the key to be available, at least one replicator in each KV's replication set must correctly store and return the KV (Figure 11.5 (b): one cell per row). For the encryption key to be securely deleted, at least one entire replication set must securely delete their KV (Figure 11.5 (c): one full row). Note that *one cell per row does not fail* is logically equivalent to the negation of *one full row does fail*, and so these properties exist in duality when the failure locations are inverted.

The $\times$ symbol is used to indicate a failure. A failure in Figure 11.5 (b) means that a KV fails to be available, either because it was erased or corrupted. This can occur as a hardware problem in the storage medium, its mismanagement, etc. A failure in Figure 11.5 (c) means that a KV fails to be securely deleted, because it was exposed, improperly deleted, or is no longer able to be deleted. This can occur because a drive is lost, becomes read-only, breaks down and is sent for repair, develops bad blocks that cannot be deleted, etc.

As the size of the key class increases, secure deletability increases while availability decreases. Similarly, increasing the size of the replication set increases availability and decreases secure deletability (assuming an independent chance of failures across nodes). The client selects the key class—both the number of entries and the pools for each entry—permitting the client to select a desirable trade-off.

An XOR-based encoding with x operands and c copies of each operand yields a secret encoding capable of tolerating, in the worst case, $x-1$ secure-deletion failures and $c-1$ availability failures. Observe that the expected number of tolerable failures improves over the worst case assuming that failures occur at random. It incurs a storage overhead of $x \cdot c$ KVs per encryption key and $c-1$ stored bytes are not read for every byte read during reconstruction. The replication of KVs permits the nodes to among themselves automatically recover from $c-1$ missing KVs and $\lceil \frac{c}{2} \rceil - 1$ corrupted KVs.

11.6.3.1 Polynomial-Interpolation Encoding

Polynomial-interpolation encoding, also called Shamir secret sharing [94], divides an encryption key into n shares (KVs) such that any k-sized subset of them can reconstruct the secret. This is done by choosing a random polynomial f with degree $k-1$ and coefficients in GF(2^{κ}) (recall κ is a security parameter corresponding to the length of binary strings used for encryption keys). The encryption key is encoded as the polynomial's constant term, $f(0)$, and the n KVs (i.e., secret shares) are $f(p_1),\ldots,f(p_n)$, where p_i is an arbitrary non-zero coordinate associated to the keystore node that stores the KV. Given k distinct points, one can interpolate the unique polynomial of degree $k-1$ and thus evaluate the constant term. Importantly, $k-1$ points provide no information about any other point, thus preserving the security of the encryption key.

The client must explicitly store at least $n-k$ coordinates with the keystore nodes. The first k coordinates can be randomly assigned from keystore nodes and the encryption key determined by interpolation. Once the polynomial is fixed, however, the remaining points $n-k$ are deterministically computed by the client and provided to the remaining keystore nodes. Availability guarantees are not provided until this completes. The client generates the remaining shares to ensure that no keystore node is aware of the polynomial (and thus the KV): doing so would violate our requirement that only the client is aware of the ultimate encryption key.

Assuming that KV shares are not replicated, a k-out-of-n secret-sharing scheme permits $n-k$ availability failures and $k-1$ secure-deletability failures. By selecting n and k so that $k-1 \simeq n-k$, the encoding is equally effective at both types of failures.

Without replication of KVs, detecting and correcting missing/corrupted KVs requires the client to retrieve all coordinates and interpolate the polynomial. To permit the nodes to verify that their shares are accurate (without in fact learning the secret) additional auxiliary information must be stored. Chor et al. [115] first introduced the notion of a *verifiable secret-sharing* scheme that permits detection of faulty shares with $O(k)$ per-share extra information. Feldman provides a non-interactive approach that adds k KV-sized values per share [116]. Herzberg et al. [22] provide an approach that permit erasure and corruption recovery of shares at the cost of storing $n-1$ KV-sized values per share.

11.6.3.2 Encoding Comparison

Table 11.4 Comparison of XOR and polynomial-interpolation encoding.

	XOR		poly/verify/recover[a]	
	3x3	**4x4**	**3-of-5**	**4-of-7**
worst-case deletion failures	2	3	2	3
worst-case availability failures	2	3	2	3
average case[b] deletion failures	3	5	2	3
average case[b] availability failures	5	10	2	3
erasure corrections	2	3	0/0/2	0/0/3
corrupted corrections	1	1	0/0/2	0/0/3
required availability	$\frac{1}{3}$	$\frac{1}{4}$	$\frac{3}{5}$	$\frac{4}{7}$
storage overhead (KVs)	9	16	5/20/25	7/35/49
assign traffic (KVs)[c]	6	8	5	7
read traffic (KVs)	3	4	3	4
discard traffic (KVs)[c]	6	8	5	7

[a] values in the form A/B/C report Shamir secret sharing as A, Feldman's verifiable sharing as B, and Herzberg et al.'s proactive system as C.
[b] computed in simulation.
[c] assumes a quorum of two without over reporting.

Table 11.4 presents the attributes of different example encodings. It shows the number of secure-deletion failures and availability failures it tolerates for both the worst and average cases, error-correction attributes, the server availability ratio required when reading, and the overhead in storing, assigning, reading, and discarding encryption keys measured in KVs. We compute the average-case failures in simulation by building the encoding and repeatedly counting the number of random failures it tolerates before the key become irrecoverable (for availability) and indelible (for deletion); we take the median over 100,000 trials. We only do this for the XOR encoding because the average case equals the worst case for the polynomial encoding. We report storage overhead and error-correction capabilities for the polynomial encoding using Shamir's, Feldman's, and Herzberg et al.'s secret-sharing schemes. Note that the latter two increase storage overhead but this data does *not* need to be securely deleted.

11.6.4 Security Analysis

We now analyze our system's security. We begin by showing how it provides secure deletion against our adversary by inspection of the adversary's capabilities. We then present other security considerations. In this section, we use the term *data lifetime* to refer to the data's lifetime expanded in both directions by the upper bound on existential and deletion latency.

Secure data deletion requires that an adversary who performs coercive attacks at all times outside the data's lifetime and non-coercive attacks during its lifetime is unable to recover the data. By encrypting stored data and computationally bounding our adversary, this corresponds to the adversary being unable to recover the KVs that create the data item's encryption key.

The keystore properties ensure that the adversary's coercive attacks do not reveal non-valid data. **P4** ensures that coercive attacks outside the lifetime of data do not recover the KVs, while **P1** ensures that the KVs are uniquely assigned and unguessable. Now we show how we provide security despite the adversary's four non-coercive attacks, which can be performed at any time. Section 11.2 in this chapter describes these attacks in detail.

- **A1** (accesses content store). This provides no information because all data written to the content store is encrypted. ATs can only be decrypted by performing a coercive attack, but non-valid data is securely deleted by the keystore.
- **A2** (acts as a keystore client). The adversary cannot guess ATs (**P4**) and so is unable to read or discard KVs for which it was not assigned. The adversary is able to deny service by repeatedly calling `assign` to obtain KVs.
- **A3** (operates a keystore node). The adversary can obtain a fraction of the KVs in the system. We use complementary key pooling to ensure that no single node is capable of deriving the key using local information. Certificates that bind keystore nodes to the pools they serve also prevent nodes from joining replicating sets for KVs from complementary pools and from performing Sybil attacks.
- **A4** (mounts network attacks). All communication is done over TLS-secured forward-secure channels: eavesdropping or modification is not possible. An adversary may mount a denial-of-service attack on the communications network.

Denial-of-Service Attacks.

Denial-of-service (DoS) attacks remain an open problem in computer security. Our adversarial model assumes that a long-term DoS of all keystore nodes is not possible. There are precautions to take that mitigate the possible damage.

The first DoS attack occurs when the adversary uses **A2** to assign all possible KVs and thus consume all keystore nodes' free storage capacity. If clients are authenticated and known, then simple accounting can tally the difference between assigned and discarded KVs; either charging the client for the storage or placing limits on the number of assignable KPs. When keystore nodes operate as a public service then a micropayment mechanism may be used when assigning KVs.

The second DoS attack occurs when the adversary, using **A4**, interferes with network communications. While this inconveniences users' storing and retrieving of data, it also has implications for secure deletion. In particular, it increases the deletion latency beyond the key pool's promised upper bound in two ways: (i) by preventing the replication set from synchronization; (ii) by preventing the client from being able to contact keystore nodes to discard KPs. Note that neither affects existential latency: the assignable set of KPs excludes ones that are not recently synchronized. To mitigate (i), keystore nodes that observe a long delay in synchronization may securely delete all discarded KVs without updating the rest of the page. If the client informed the entire replication set before the DoS began then this achieves the secure deletion of this data. Mitigating (ii) is more troublesome; the distribution of keystore nodes should make it difficult for the adversary to inhibit all communication.

Session Keys and Entropy Pools.

Bounded existential and deletion latencies require that compromising the keystore nodes' states does not reveal valid KVs outside their valid lifetime. When refining the design to an implementation, it is important to securely delete any data that can violate this property. Perfect forward secrecy is often implemented with periodic key renegotiation; random numbers are typically generated by drawing from a hardware entropy pool. An adversary with knowledge of the entropy pool may be able to predict session keys; an adversary with knowledge of a session key can determine the cryptographic random number generator used to generate new KVs. To guarantee latency bounds, session keys and the entropy pool's contents must be periodically refreshed and securely deleted after use.

Figure 11.6 shows how the adversary's knowledge of the entropy pool and session keys effects the compromise of KVs. While our design expects that the adversary's single attack only obtains KVs valid at that time, knowledge of the entropy pool allows future session keys to be predicted and therefore effect a compromise of future KVs. Knowledge of the current session key allows the adversary to decrypt previously collected network traffic and determine earlier KVs. We must renegotiate session keys and clear the entropy pool with the same timeliness as the shortest key pool synchronization.

Key Value Freshness.

The client can confirm KV freshness when over-reporting an assign: the KV is fresh if any replicator believes that the KP's state is **U**. This is the case unless the assigner led a synchronization before the client is able to report. Further protection can be added by having each node maintain a set of recently assigned KPs.

Multiple Assigns.

We assume that assigners do not intentionally multiply assign KPs, and unintentional failures due to lost state are avoided by first resynchronizing before assigning. We nevertheless consider how to avoid malicious assigners.

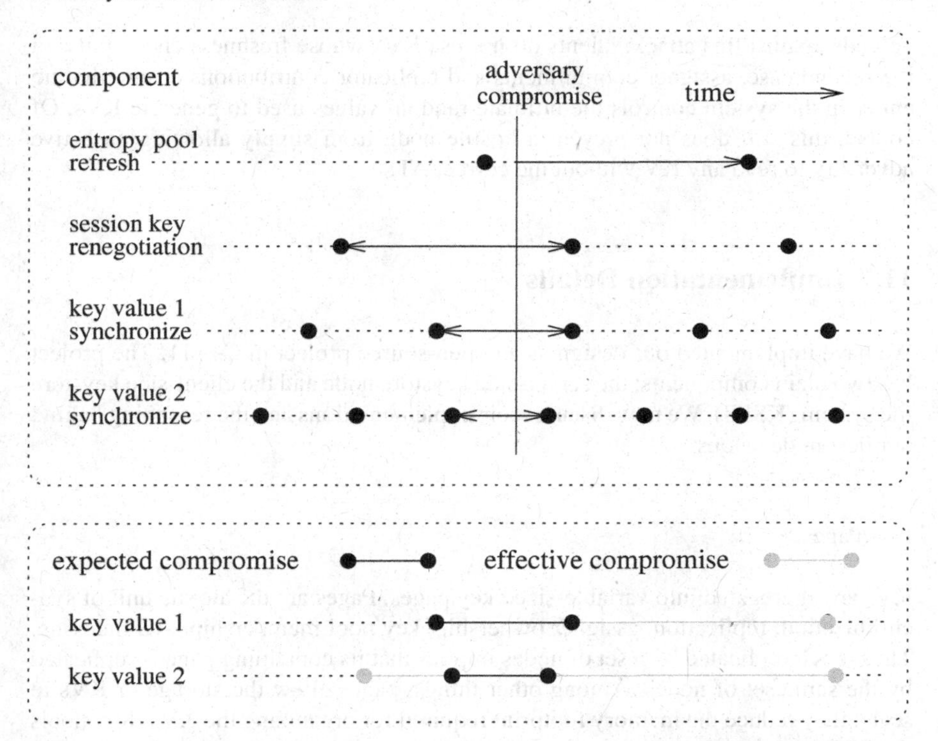

Fig. 11.6 Effect of entropy pool and session key disclosure on the compromisability of KVs. The expected compromise is indicated in black while the effective compromise due to session keys and entropy pools is indicated in grey.

KV freshness ensures that a malicious assigner cannot doubly assign KVs that were assigned before the most-recent synchronization. Assigners can, however, assign a **U** KP to multiple clients before synchronization. While client IDs for **A** KPs can be compared during synchronization, this increases the communication cost between nodes.

Covert Disclosure.

A possible attack is a keystore node disclosing future KVs to an adversary well in advance of their use, thus avoiding suspicious behaviour at the time the KV is assigned. Observe that this is equivalent to effecting a large existential latency for data.

A node can increase existential latency in two ways: (i) by fraudulently assigning all stored KVs among the nodes in advance, reporting the KVs to the adversary, and then only later actually assigning them to clients; (ii) by generating new KVs predictably, e.g., with predictable random seeds. In the first case, value freshness defends against this attack: clients do not use KVs whose freshness check fails. In the second case, assigner commitments and replicator contributions ensure that no entity in the system controls the ultimate random values used to generate KVs. Of course, this still does not prevent a hostile node from simply allowing an active adversary to read any KV without the correct AT.

11.7 Implementation Details

We have implemented our design as an open-source project in C++11. The project has two main components: the server-side keystore node and the client-side keystore file system (KSFS). We now discuss their implementations and the reasoning behind our design decisions.

Key Pages.

KVs are aggregated into variable-sized key pages. Pages are the atomic unit of synchronization, replication, assigner ownership, key pool membership, and indexing. Thus, a KP replicated by a set of nodes implies that its containing page is replicated by the same set of nodes. Among other things, pages allow the storage of KVs to scale: they reduce the memory footprint required for organizing the KVs. Key pools therefore consist of a set of key pages.

Key pools may optionally offer store pages, which are used to store KVs explicitly provided by the client. This is used for encodings with algebraic structure such as polynomial-interpolation encoding. The client calls the keystore node's `store` function, providing the KV to be stored; the node returns the KP for later retrieval.

Key-Page State and Synchronization.

Each page stores two bits for its position's state. The first bit (the state bit) for each position is **true** if the position's state is **A** and **false** if it is not **A** (i.e., either **U** or **D**). The second bit (the delta bit) is **true** if and only if the state bit is different from the last synchronized consensus.

Synchronization is performed for an entire page. Instead of providing the full state, nodes provide their LZO-compressed [117] delta bit vector. The KPs for KVs that are both assigned and discarded since the most-recent synchronization are specifically listed after the delta vector for efficiency. The assigner also provides the state vector as it was after the previous synchronization, which the replicators check against their own. Each node in the replication set then updates the KP's state according to the `vote-state` algorithm. Assigned positions whose local discard was rejected by consensus retain the KV and the **D** state with the delta bit set.

The page's **U** and **D** KVs are replaced with new values taken from a cryptographic random number generator [118]. This generator is seeded by the bitwise XOR of all the random value contributions from each node in the replication set. Note that this refinement replaces the i.i.d. distribution with a pseudo-random distribution seeded by randomness.

Key-Page Cache.

The key-page cache manages access to pages, keeping some keys in memory and loading the others from long-term storage when needed. Any page use requires its availability in the cache. Each pool has its own page cache. The policy defines the minimum size of the memory used to cache pages, thereby improving the expected performance.

Depending on how pages are stored, loading them from long-term storage may require complicated operations, including determining the page's key and decrypting its contents. All page metadata, including each KP's state, is loaded alongside. The caching strategy keeps assignable pages ready in the cache and evicts the replicated pages based on a least-recently used metric, which has low overhead and high performance. Pages with local modifications are first written to long-term storage before eviction.

Key-Page Locking.

For thread-safety, key pages are locked when performing mutable operations, e.g., assigning and deleting keys. Once a page is locked, the time required to handle an assign or delete request is extremely brief. Originally, when assigning a key, we polled for an unlocked page if one was available instead of waiting on a locked page; testing, however, revealed that at high load it is more efficient (and at low load it made no difference) simply to pick a random assigner page and wait for the lock.

While KV operations (i.e., assign, read, and discard) take constant time, a few key-page operations take time proportional to the page size. These are marshalling a page for replication, preparing the pull message, and updating the page with the new version. The synchronization protocol does not affect performance for KV operations: from the time a page's pull message is generated until the time the page is finished synchronizing, the page is replaced with an update wrapper. This update wrapper caches all discard requests itself, rejects any assign requests (nodes maintain multiple assigner pages per pool to account for locked pages), and handles read requests without locking the page. After updating the key page, the update wrapper's modifications (i.e., deletions) are replayed and the wrapper removed. Therefore, the only operation that may affect the latency of KV operations is marshaling a key page for replication. Since this operation primarily happens before a page is ready for use and only occasionally when a replicator leaves the network, we conclude that page lock time is not a concern.

Key Builder.

The *key builder* is a client-side service that processes key classes and key recipes, performs the appropriate calls to keystore nodes, and returns the actual encryption keys. The key builder's functions are the following: `read_key`, which turns a key recipe into an encryption key by reading the referenced KVs and assembling the key; `assign_key`, which takes a key class and returns both an encryption key and its corresponding key recipe by getting new KVs for each key pool in the class; `discard_key`, which takes a key recipe and discards each constituent key position. Each of these functions executes in parallel with respect to keystore operations: reading a key recipe of three KVs, for example, spawns three simultaneous read requests. When KVs are replicated over multiple keystore nodes, the key builder selects among them until the request is satisfied. These functions take the further step of decrypting and encrypting the key recipes using the client's password-derived key.

Polynomial Interpolation.

We implement k-out-of-n polynomial interpolation using a refined key builder that modifies the `assign_key`, `read_key`, and `discard_key` functions. Polynomial key classes list the set of key pools in which shares can be stored, as well as the minimum number of shares for reconstruction (k) and the total number of shares (n). For example, there can be ten different pools, any seven of which can be used for a 3-out-of-7 secret-sharing encoding. We use Shoup's high-performance number theory library `libntl` [119], which we compiled to support `libgf2x`'s fast arithmetic over binary finite fields [120]. We use the field GF(2^{128}) generated with the irreducible polynomial $1+x+x^2+x^7+x^{128}$.

The `assign_key` function randomly selects n key pools, say $p_1,\dots,p_n$. Each key pool p_i has a public domain coordinate $x_i \neq 0$ and the share is the evaluation of the polynomial at that coordinate. The function generates k random field elements $y_1,\dots,y_k$ and interpolates the polynomial f over the points $(x_1,y_1),\dots,(x_k,y_k)$. The remaining shares $y_i = f(x_i)$ are computed for $k < i \leq n$. The shares y_i are disbursed in parallel among the corresponding keystore nodes. Errors are handled by evicting the key pool from the composition and selecting a new one. Finally, the function returns the actual encryption key by evaluating $f(0)$ as well as the corresponding key recipe with metadata indicating the polynomial's degree. The `read_key` function takes the key recipe and randomly selects k keystore nodes to retrieve the shares, interpolate the polynomial, and return the encryption key. Errors are resolved by later retrieving other nodes' shares. The `discard_key` function behaves as the normal key builder, discarding each constituent KV.

Keystore File System.

The keystore file system (KSFS) is a file system that provides secure deletion for its data using the solution proposed in this work. KSFS is a FUSE-based file system that stacks on top of an existing file system, e.g., a file system that proxies access to the content store.

KSFS encrypts each data block with its own unique key. The recipe is stored along with the encrypted block in the file or as a separate related file. Moreover, each file is optionally given its own master key. This key is assigned from the nodes in the same manner as other keys, though it may use a different key class. The master key is used to decrypt the block keys used by the file, therefore, the block key retrieved through the recipe is interpreted as an encrypted key. When the file master key is securely deleted, then all the content in the file is also securely deleted even if some per-block keys remain. This provides another defense for compromised keys as well as a greater efficiency when securely deleting an entire file.

Only a few features were added to the FUSE file system to implement our design: initialize a key builder service, store and retrieve key recipes alongside encrypted data, and perform the relevant cryptographic operations in the I/O datapath. The KSFS is implemented by making the following changes to the FUSE file system:

- `create`: obtain a file master key and store its file recipe at the beginning of the file; allocate auxiliary state.
- `open`: read the file master key and build its key; allocate auxiliary state for file.
- `truncate`: delete the recipes for all truncated blocks; re-encrypt the file's tail if the truncation is unaligned with the file system block size.
- `unlink`: delete all recipes for the file including the master key.
- `flush`: flush file auxiliary state if necessary.
- `read`: read a part of the file as well as the key recipes for that part; build the corresponding keys and decrypt the data.
- `write`: obtain new keys from the key builder, encrypt the data, and write the encrypted data and key recipes to the file.
- `statfs/getattr`: correct file sizes to account for recipes.

In reality, the read and write functions are more complicated. Since the file system's block size is the granularity of encryption, writing data that replaces an existing part of a file—but not a complete block—requires reading and decrypting that part of the file, making the change, re-encrypting the block with a new key, and updating the full block and key recipe. Consequently, the encryption block size affects system performance because all operations must occur with that block size as the minimum unit of I/O.

To improve performance for repeated small reads or writes within a block, a one-item cache is used to store the current working data block. When a read or write requires a particular block, the cache is first checked for its presence. If absent, the current block is written to the content store (if modified in cache) and replaced with the desired block. All reads and writes are performed only on the cached entry.

Keystore Communication.

We use RPC calls for all keystore communication. Each keystore node is assigned an IP address and port to run an RPC server. Our implementation uses `xmlrpc` [121] for the RPC server and `stunnel` [122] to expose it as an `https` [123] service using OpenSSL [124], thus permitting TLS connections to persist over multiple RPC calls. Consequently, the IP address of the RPC server is only accessible locally and used for forwarding traffic from the publicly available TLS-secured RPC server. Additionally, the IP, port, and public key of some initial *bootstrapper* nodes' RPC servers are given to each new keystore node that joins. On initialization, the node queries one of the bootstrapper nodes to request known peer nodes. The nodes that are provided are further queried until the relevant network view is known. When querying for nodes, the list of pools served by the node is also provided. The node ignores any other nodes with which they do not share a pool. The KSFS also uses RPC to communicate with the nodes to request keys, etc. The KSFS is also bootstrapped in the same way. Since the KSFS defines the classes it uses, it can also determine the set of pools that interest it.

11.8 Experimental Validation

We validate our design by experimenting with our implementation. We focus on the inter-keystore-node communication cost during synchronization and the keystore's latency and throughput for satisfying clients' requests. We perform our experiments using 14 ThinkCenter M91p desktops in a local area network, one of which acts as the content store and the remainder divided into keystore nodes and clients. We envision deployment settings where keystore nodes serving the same key pools are within the same data centre and where the client's latency to the keystore nodes is comparable to the latency to the content store.

Inter-node Communication.

We measure the amount of communication in our system by summing the size of outgoing inter-node RPC calls. Communication mostly occurs when nodes synchronize pages. Our goal is to verify that compressing synchronization messages and seeding random number generators for creating KVs result in low communication overhead.

We do this by measuring the average inter-node communication in a distributed system of 6 or 12 computers running keystore nodes. We use a baseline configuration that consists of 6 nodes each assigning from 1,000 16 KiB pages replicated by 2 nodes and synchronized every 10 minutes. The other configurations are generated by taking the baseline and changing one parameter.

We run the system for 90 minutes and measure the average inter-node communication over this time with and without data compression. Note that *compressed* reflects the best case as no state changes occur; *uncompressed* is the worst case where enough random state changes occur to render it uncompressable. We further compute the number of unique key positions provided by this configuration. Table 11.5 presents our results. Note that the communication is measured as the *per-node outgoing* communication—the total system communication is scaled by the number of nodes.

Table 11.5 Average inter-node communication with 95% confidence. Each row has different parameter configurations. The `baseline` configuration has the following values: 6 nodes each assigning from 1,000 16 KiB pages replicated by 2 nodes and synchronized every 10 minutes. The remaining measurements vary from the baseline by the identified parameter.

configuration	key values (millions)	communication (B/s) compressed	uncompressed
baseline	6.1	4380 ± 24	7285 ± 3
2000 pages	12.3	8597 ± 24	14281 ± 11
5 minute clock	6.1	8704 ± 22	14460 ± 7
64 KiB pages	24.6	4221 ± 14	13089 ± 6
1 replicator	6.1	1856 ± 23	3053 ± 26
5 replicators	6.1	8738 ± 22	14513 ± 3
12 nodes	12.3	4638 ± 23	7724 ± 23

Table 11.5 provides evidence to support our hypothesis on the utility of aggregating KVs into key pages: given a fixed client load, pages can be synchronized independent of page size. Note that while our tests had all pages synchronized, in practice keystore nodes only need to synchronize pages with local changes along with sufficient fresh assigner pages to provide KVs. Unchanged pages can delay their synchronization, though a mechanism that allows replicators to initiate synchronization when discards occur is necessary.

The size of the replication set affects the communication complexity because messages are passed among more nodes. Communication scales quadratically to the replication set.

Increasing the total number of nodes does not affect communication complexity. The additional overhead in discovering nodes and negotiating secure channels is therefore significantly less than the cost of performing synchronization. The communication complexity scales with the number of pages being synchronized and inversely with the period between synchronizations.

Latency and Throughput.

Table 11.6 presents latency for basic key operations. Internal latency is the machine-local processing time required to handle keystore requests and does not include net-

Table 11.6 Keystore operation 95th percentile latency with 95% confidence. The first table uses the XOR encoding. Internal latency is a single KV operation's processing time within the node, i.e., after the RPC call has been received and until the response is prepared. External latency is the client's observed time for a single KV operation, including local network latency. Key building latency is the client's observed time for performing an operation for an entire key recipe. The second table uses polynomial interpolation. It includes generating a new polynomial's shares and interpolating a polynomial.

	latency (μs)		key building latency (μs)		
operation	**internal**	**external**	**2 parts**	**4 parts**	**6 parts**
assign	24 ± 1.0	726 ± 7	1034 ± 21	1115 ± 36	1414 ± 39
read	16 ± 0.7	725 ± 8	1102 ± 27	1181 ± 10	1416 ± 30
delete	28 ± 1.0	719 ± 9	1070 ± 34	1198 ± 11	1423 ± 35

recipe	**polynomial-interpolation key-building latency (μs)**			
operation	**2-out-of-3**	**3-out-of-5**	**4-out-of-7**	**5-out-of-9**
generate	135 ± 1.5	204 ± 6.3	307 ± 6.4	423 ± 6.8
interpolate	25 ± 0.9	31 ± 0.8	38 ± 0.8	47 ± 0.8
assign	1135 ± 6.9	1331 ± 6.4	1532 ± 32.5	1666 ± 19.5
read	1072 ± 4.7	1090 ± 7.2	1136 ± 7.2	1279 ± 10.9
discard	1059 ± 13.6	1239 ± 10.4	1441 ± 17.2	1630 ± 16.6

work latency; external latency is the client's observed latency for request processing including HTTPS tunneling and local network latency. Key-building latency is the time taken to assign, read, or discard an entire key. We measure key-building latency for the XOR encoding using different recipe sizes: two, four, and six. We measure it for the polynomial-interpolation encoding for 2-out-of-3, 3-out-of-5, 4-out-of-7, and 5-out-of-9 secret sharing. Measurements are computed by averaging six 95th-percentile observations; the true mean is within the interval 19 times of out 20.

Latency measurements are stable across the parameters and workloads we tested. In particular, all page sizes from 4 KiB to 128 KiB exhibit the same latencies. The internal latency of read operations is smaller because both assign and discard events are logged for crash recovery and the journal synchronized before returning to the client.

The RPC library, the HTTPS server, and the network add significant overhead to the latency of requests, from a couple dozen microseconds to a millisecond. The use of HTTP-based transport is not optimal for binary data because it must be encoded into base64. Replacing the RPC library should therefore yield better performance. The ping between machines is 300 μs and the loopback device (for communications between `stunnel` and the HTTP server) has a delay of 10 μs.

The latency for recipe operations scale sub-linearly with the recipe size, even at the 95th percentile. Assigning a six-part recipe, for example, only takes 40% more time than a two-part recipe. Complementary key pooling allows the per-position operations to be dispatched in parallel and so the latency is chiefly waiting for the slowest response.

To measure the throughput of the keystore, we created a client that issues random requests with high frequency. In order to actually consume the entire capacity of the keystore, however, we used a network of two nodes serving a single pool and used the other 10 computers to issue requests. With this setup, we achieved a throughput of 28 ± 0.1 kilo-operations per seconds, with each node handling approximately 14 kilo-operations per second. Despite the high load, the key operations' latencies during throughput measurements still fell within the mean's confidence interval. Latency at high load is therefore competitive with low load.

11.9 Conclusions

We developed robust key storage for secure data deletion using both a persistent storage medium and a securely deleting storage medium. The proposed system can handle the partial failure of the securely deleting storage, either by failing to delete data or failing to store data. Our design distributes the securely deleting storage medium over many nodes and allows the client to select from key values with different storage and deletion policies. We consider an encoding scheme that balances both data secure deletability and data availability. We implement our design and analyze its performance, observing that it has a high service rate and can synchronize many securely deletable key values with a low communication complexity.

11.10 Research Questions

- What is the best way to reason about the XOR-based encoding when, unlike polynomial-encoding, failures are not fungible among the nodes?
- What is a good formalization of a keystore as a secure-deletion primitive component which can be easily refined for specific deployments and used to formally reason about secure deletion?
- Is it possible to detect the silent compromise of the user's password (e.g., because it was guessed) through the use of fake recipes—say, *honey recipes*—that are indistinguishable from legitimate ones but which signal to the keystore nodes that the user's password has been compromised?
- Our use of access tokens is motivated by the desire to keep the synchronization overhead constant per-key-page instead of per-key-value. Is access control based on client IDs possible without over-reporting assigns to all replicators?
- What is the throughput and performance of such a system on a real-world data-centre-based cloud storage system?

11.11 Practitioner's Notes

- Our failure model assumes an independent chance of failure, as is common. Failures of secure deletability and partial adversarial compromise, however, may not in fact be independent and this may warrant a more appropriate model.
- Currently there is no budget allocations for clients when assigning keys. Real deployments will either trust clients to request keys only when needed and always delete, or have a consistency mechanism that allows the discovery of orphaned assigns as well as determine which KVs each client has requested.
- Key pools can also impose performance requirements on the nodes, e.g., their latency and throughput in satisfying requests. Pools may also place geographic restrictions on where the nodes may operate, e.g., limited to a particular set of countries.

Part IV
Conclusions

Chapter 12
Conclusion and Future Work

With this final chapter we conclude our book. We begin by summarizing our contributions and comparing them with related work using the framework we developed in Chapter 2. We then present a short discussion on some related and complementary lines of research that fall outside our scope. We follow this by unanswered questions and directions for future work. We finish by drawing our conclusions.

12.1 Summary of Contributions

In Chapter 2, we presented related work on secure deletion and compared their environmental assumptions and behavioural properties in Table 2.3. Related work for flash and cloud storage were presented in subsequent chapters as well as our own contributions: user-level deletion, DNEFS/UBIFSec, securely deleting B-Trees, and distributed keystores. Our work was generally designed to defeat a strong adversary: the computationally bounded unpredictable multiple-access coercive adversary. Our solutions are further designed to provide efficient secure deletion at a fine (per-data-item) granularity. Table 12.1 compares all this work. Building on Table 2.3, it includes flash- and cloud-related work as well as our own.

12.2 Related and Complementary Research

We present a focused study on the topic of secure data deletion. There is a great wealth of research done on related and complementary topics, however, that lie out of scope for this book. This chapter highlights a variety of these topics and relevant research articles. Some of these are orthogonal to our work; others are higher-level concepts that build on the sort of secure data deletion primitives we focus on in this book.

J. Reardon, *Secure Data Deletion*, Information Security and Cryptography,
DOI 10.1007/978-3-319-28778-2_12

Table 12.1 Full spectrum of secure-deletion solutions. This table expands on Table 2.3 to include related work for cloud and remote storage, as well as the solutions proposed in this book.

Solution Name	Target Adversary	Integration	Granularity	Scope
overwrite [35,36,48]	unbounded coercive	user-level[a]	per-file	targeted
fill [37,38,57]	unbounded coercive	user-level	per-data-item	untargeted
NIST clear [11]	internal repurposing	varies	per-medium	untargeted
NIST purge [11]	external repurposing	varies	per-medium	untargeted
NIST destroy [11]	advanced forensic	physical	per-medium	untargeted
ATA secure erase [28]	external repurposing	controller	per-medium	untargeted
renaming [46]	unbounded coercive	kernel[a]	per-data-item	targeted
ext2 sec del [16]	unbounded coercive	kernel[a]	per-data-item	targeted
ext3 basic [46]	unbounded coercive	kernel[a]	per-data-item	targeted
ext3 comprehensive [46]	unbounded coercive	kernel[a]	per-data-item	targeted
purgefs [49]	unbounded coercive	kernel[a]	per-data-item	targeted
ext3cow sec del [51]	bounded coercive	kernel[a]	per-data-item	untargeted
compaction	unbounded coercive	kernel	per-data-item	untargeted
batched compaction	unbounded coercive	kernel	per-data-item	untargeted
per-file encryption [77]	bounded coercive	kernel	per-file	targeted
scrubbing [26]	unbounded coercive	kernel[a]	per-data-item	untargeted
flash SAFE [7]	external repurposing	controller	per-medium	untargeted
purging Chap. 5	unbounded coercive	user-level	per-data-item	untargeted
ballooning Chap. 5	unbounded coercive	user-level	per-data-item	untargeted
hybrid Chap. 5	unbounded coercive	user-level	per-data-item	untargeted
DNEFS Chap. 6	bounded coercive	kernel	per-data-item	untargeted
revocable backup [5]	bounded coercive	composed[b]	per-file[c]	targeted
forget secret [92]	bounded coercive	composed[b]	per-data-item	targeted
Ephemerizer [2]	bounded coercive	composed[b]	per-data-item[de]	targeted
Ephem. time based [9]	bounded coercive	kernel	per-data-item[e]	targeted
Ephem. on demand [9]	bounded coercive	kernel	per-file	targeted
Ephem. classes [9]	bounded coercive	kernel	per-data-item[f]	targeted
porter devices [10]	bounded coercive	composed[b]	per-data-item[de]	targeted
Vanish [13]	bounded coercive	composed[b]	per-data-item[d]	targeted
Fade [3]	bounded coercive	composed[b]	per-data-item[f]	targeted
policy-based [93]	bounded coercive	composed[b]	per-data-item[f]	targeted
B-Tree Chap. 10	bounded coercive	composed[b]	per-data-item	targeted
keystore Chap. 11	bounded coercive	composed[b]	per-data-item	targeted

[a] Assumes interface performs in-place updates.
[b] Non-standard interface, assumes two storage media: one securely deleting and one persistent.
[c] Works at the granularity of a backup, which is composed of arbitrary files.
[d] Data items are messages communicated between two peers.
[e] Data items' lifetimes are known in advance.
[f] Data items are grouped into classes and deleted simultaneously.

Solution Name	Lifetime	Latency	Efficiency
overwrite [35,36,48]	unchanged	immediate	number of overwrites
fill [37,38,57]	unchanged	immediate	inverse to medium size
NIST clear [11]	varies	immediate	varies with medium type
NIST purge [11]	varies	immediate	less efficient than clearing
NIST destroy [11]	destroyed	immediate	varies with medium type
ATA secure erase [28]	unchanged	immediate	inverse to medium size
renaming [46]	unchanged	immediate	truncations copy the file
ext2 sec del [16]	unchanged	immediate	batches to minimize seek
ext3 basic [46]	unchanged	immediate	batches to minimize seek
ext3 comprehensive [46]	unchanged	immediate	slower then ext3 basic
purgefs [49]	unchanged	immediate	number of overwrites
ext3cow sec del [51]	unchanged	immediate	deletes multiple versions
compaction	some wear	immediate	inefficient, lots of copying
batched compaction	some wear	periodic	no worse than compaction
per-file encryption [77]	some wear	immediate	one erasure at file deletion
scrubbing [26]	unchanged	immediate	varies with memory type
flash SAFE [7]	some wear	immediate	inverse to medium size
purging Chap. 5	some wear	immediate	depends on medium size
ballooning Chap. 5	variable wear	no guarantee	tradeoff with deletion
hybrid Chap. 5	variable wear	periodic	periodic duration trade-off
DNEFS Chap. 6	little wear	periodic	periodic duration trade-off
revocable backup [5]	varies[a]	periodic	re-encrypt all backup keys
forget secret [92]	varies[a]	immediate	re-encrypt tree path
Ephemerizer [2]	varies[a]	periodic[b]	one key per expiration time
Ephem. time based [9]	varies[a]	periodic[b]	one key per expiration time
Ephem. on demand [9]	varies[a]	periodic	re-encrypt all file keys
Ephem. classes [9]	varies[a]	immediate[c]	must delete all data in a class
porter devices [10]	varies[a]	immediate	uses public key crypto
Vanish [13]	varies[a]	no guarantee	requires DHT access
Fade [3]	varies[a]	periodic[c]	only delete policy atoms
policy-based [93]	varies[a]	periodic[c]	only delete policy atoms
B-Tree Chap. 10	varies[a]	periodic	re-encrypt tree path
keystore Chap. 11	varies[a]	periodic	depends on key encoding

[a] Wear depends on how the securely deleting storage medium is implemented.
[b] Deletion time is selected in advance from a finite set of possibilities.
[c] Deletion is based on logical expressions from a finite number of terms.

12.2.1 Information Deletion

In our work we assumed that the user was able to identify the data items that need to be deleted. In general, however, the user may only know that some *information* must be deleted, e.g., a report, and may not reliably identify all data items that constitute or are relevant to the information that is to be deleted, e.g., spreadsheets, images, earlier drafts. Information flow techniques, such as taint analysis, can be used to identify related content.

Ritzdorf et al. [125] introduce a system to help the user identify related content for the purposes of secure deletion. Their system uses a temporal-spatial metric for data item relatedness: whether pairs of data items are commonly accessed together. As their metric to efficiently estimate relatedness, they use whether it is colocated in the operating system's cache. This metric is continually updated to strengthen the relatedness. When the user deletes a file, they are presented with a list of other files as candidates for deletion at the user's discretion.

12.2.2 File Carving

In our work we assumed that an adversary capable of recovering all pieces of a data item can recover the data item itself. In practice, however, this can be far from trivial. Simply having an unordered sequence of binary strings that correspond to pieces of many different files does not admit easy reconstruction.

Solving this problem is the domain of *file carving techniques* [126–128]. These rely on statistical inferences of the kind of data contents and which blocks are semantically adjacent. For example, a reconstructed text file will form words and sentences [129]; a reconstructed image file will have similar coloured pixels across the arbitrary block divides [130]. More formally, the task of file carving is to efficiently compute a statistically likely permutation of the file system's blocks.

12.2.3 Steganographic and Deniable Storage

In our work we assumed that the use of a secure-deletion tool as well as evidence that data was at some point stored and deleted was inconsequential. This may not always be the case: some users may desire to hide any evidence data was stored and deleted on a storage medium, or even hide that a secure-deletion tool was used to delete data.

A steganographic file system is one that stores a small amount of data innocuously within the file system [131]. A log-structured file system could, for example, store bits of data through the way files are interleaved into segments. Another opportunity for data storage is in the unused or free space of the file system. Provided that the unused space is filled with random data that is indistinguishable from encrypted data, the owner of a deniable file system can *plausibly deny* that there is no hidden data on the storage medium. Of course, the mere *use* of a deniable file system may already cast suspicion, though this can be mitigated if operating systems support deniable file systems by default or the deniable file system can be mounted as a traditional file system when accessing the non-steganographically stored data.

Skillen and Mannan design an Android-based deniable file system called Mobiflage [132]. They use a *hidden-volume* approach whereby a large chunk of the file system (25–50% in their case) is reserved a priori to contain a hidden volume within an encrypted outer file system. To an external observer, the entire file system appears to contain random data; even with the encryption key, the hidden volume looks like random data stored in the unallocated space of the file system, and the user must take care not to write data into the hidden space in the file system. The user accesses the hidden volume with a secret password but can plausibly deny its existence.

Peters et al. design a YAFFS-based implementation of a log-structured deniable file system called DEFY [133]. They use a *steganographic* approach whereby secret data is seamlessly woven throughout the file system. They leverage the widespread obsolete blocks in log-structured file systems to store steganographic data. To ensure that secret data looks indistinguishable from obsolete data, they provide fine-grained secure data deletion in a manner similar to DNEFS with keys arranged in a tree-structure. Specifically, they use an all-or-nothing transformation, similar to Peterson et al. [50], to create a *tag* for each data item that serves the similar function as the encryption keys in DNEFS: the deletion of a tag entails the secure deletion of the data associated to that tag. Instead of managing all tags separately, however, they use a hierarchical approach: directories decrypt the contained files, files decrypt the contained data blocks. Updates result in a "cascade" equivalent to a shadowing mutation and the root tag is then stored separately. The authors argue that the system is deniable because, for every operation performed on steganographic data, there is at least one sequence of operations performed only on non-steganographically stored data that produces the same state.

12.2.4 History Independence

In our work we assumed that deleting data items was sufficient to delete data. The shape of some data structures and file systems, however, vary depending on the history of operations that led to its current contents. This is natural for any data structure that has isomorphisms. For example, a binary tree of three elements has five different forms that all store the same information; inspecting the current form may leak information about the order data was added or if data was deleted. A history-independent data structure is a data structure whose form is determined by its current contents and is therefore independent of the history of operations that led to its current state. While a history-independent data structure need not have *unique* representations, the distribution over its current representation must be completely determined by its current contents and not its previous state.

While history-independent versions of popular data structures exist, Bajaj and Sion observe that simply using them at the application level does not ensure history independent of data stored in the file system [134]. They solve this with the first history-independent file system, which uses a history-independent hash table for the storage of all file data and metadata. They tune the hashing function to improve sequential localization of files under the assumption of spinning-disk storage medium (i.e., high latency for random access). They leave the design of a history-independent file system for flash-based storage as future work. We note that while reconciling history independence with log-structured storage is not straightforward, the design space increases greatly as flash memory removes the constraint of data locality.

12.2.5 *Provable Deletion*

In our work we assumed that secure deletion is executed correctly by the hardware (with the exception of Chapter 11 where we considered partial failures). In the introduction, however, we stated that one challenge of secure deletion on a digital system is that of proving to the user that the data is indeed deleted with the same degree of confidence as the chaotic output of a paper shredder. Moreover, a compromised storage medium may fake the execution of secure deletion, perhaps even reserving an area of storage for the storage of data deleted by extraordinary means or for encryption keys.

Perito and Tsudik study how a resource-constrained device can actually prove the successful secure deletion of data [12] and propose a protocol called *proofs of secure erasure*. It works by sending an uncompressable random pattern to a memory-constrained device and having it return a function on the random pattern that can only be computed with the entire pattern safely stored by the device—in their case a cryptographic message authentication code serves the purpose. Karvelas and Kiayias achieve sublinear communication relative to the storage device at the cost of weakened security guarantees [135]. Hao et al. [136] propose a system to store and securely delete tamper-resistant private keys such that when a private key is deleted, the corresponding public key is included in a signed statement that the private key is securely deleted. The user is then entitled to compensation under a legal framework if the private key is ever exposed. We observed that digital currencies may be able to formalize the contracts and thus circumvent a legal framework.

12.3 Future Work

12.3.1 *New Types of Storage Media*

Storage technology advances the state of the art in many ways: capacity, reliability, performance, and price. Secure deletion, however, is not a design requirement, and creative approaches to achieve it are usually needed after new hardware is introduced.

Though new storage media may fall into a category for which solutions are known—e.g., by permitting in-place updates, by having a write/erase granularity asymmetry, or by being persistent but augmented with a securely deleting storage medium. For example, two kinds of technology currently on the horizon are shingled magnetic recording (SMR) and heat-assisted magnetic recording (HAMR).

SMR technology increases capacity by 25% by overlapping tracks (imagine roof shingles), and writes to one track can affect overlapping ones. Managing access is required either by an obfuscating hardware controller or by special drivers in the operating system. The unique geometry of this device may warrant new approaches for efficient secure deletion.

HAMR technology uses a tiny laser to heat the area of the magnetic storage being written to so that a very weak magnetic field is capable of being a magnetically coercive force. HAMR storage promises to increase the storage density 100 fold, though consumer-level devices are currently nowhere in sight. Perhaps the high density and weak magnetic field may reintroduce the types of analog remnants of deleted data first observed by Gutmann.

12.3.2 Benchmarks for Different Storage

In our experiments for mobile storage, we collected usage data of a mobile file system from a limited sample size. This was acceptable to analyze DNEFS since it is generally independent of the file system usage: the KSA size depends only on the storage medium size, and the worst-case erasure count depends on the KSA size and the clock frequency. In our experiments for remote storage, we used the replay of our research group's revision control history to simulate a shared storage medium being used to perform periodic commits of local data.

It would be useful, however, to have more datasets describing the write and discard behaviour representative of a variety of types of mobile-phone users and cloud-storage users. Having real-world open data available to the scientific community that models cloud-storage use cases, however, would provide more confidence in the performance measurements. Critically, such benchmarks must include discarding data: it is insufficient to record when a storage location is overwritten because the data item's corresponding discard time may have occurred long earlier.

12.3.3 Secure-Deletion Data Structure Selection

Our analysis of secure deletion for persistent storage proved the security for all arborescent structures. Our B-Tree-based design was motivated by the utility of having a dynamically sized data structure and a large block size when accessing data remotely; it is inspired by the ubiquitous use of B-Trees in the niche of database and file system storage. However, there is no benchmarked comparison to show that our design is better than a simple static tree or other possible arborescent structures.

It would be useful to determine the workloads and situations which are the most amenable to our design, and how other candidates compare. In addition to comparing different workloads, it would be interesting to compare the cost of adding a securely deleting B-Tree layer on top of a file system versus integrating it into an existing (B-Tree) access structure.

12.3.4 Formalization

A notable aspect missing from this work is the formalization of secure data deletion and a mathematical model of storage media. The storage medium models and the definitions of recoverable versus irrecoverable data are used as intuitive concepts and not formalized using a mathematical description of a storage medium with a definition of what can be recovered. Cryptography was assumed to be perfect: a computationally bounded adversary cannot decrypt AES-encrypted messages without the key.

It would be useful to develop formal models of the concepts we present in this work and thereby formally prove the security of our designs. One may consider a storage medium's contents through history as a transition system trace. Secure deletion might mean that the set of possible walks—with the states during the data's lifetime redacted—is indistinguishable from the set of possible walks for a different value stored instead. Another possibility is a game-based definition where the adversary wins if it correctly chooses which among two storage media actually contained some data item. A proof of correctness for a secure-deletion solution may involve proving its behaviour is indistinguishable from one of our idealized SECDEL models. In the process of formalizing and determining the best way of modelling storage media for secure deletion, useful new concepts or perspectives may be discovered.

12.3.5 DNEFS for FTLs

DNEFS efficiently solves the problem of secure data deletion for flash memory when accessed with a flash file system. However, flash file systems are less common than flash translation layers (FTL). This is because an FTL allows the storage medium to contain a traditional block-based file system such as FAT; these file systems are more widely supported among operating systems. It would therefore be useful to integrate DNEFS into a security-enhanced FTL so that secure deletion can be provided for these users as well.

While FTLs vary in implementation, many of which are not publicly available, in principle DNEFS can be integrated with FTLs in the following way. All file system data is encrypted before being written to flash, and decrypted whenever it is read. A key storage area is reserved on the flash memory to store keys, and key positions are assigned to data. The FTL's in-memory logical remapping of sectors to flash addresses stores alongside a reference to a key position. The FTL mechanism that rebuilds its logical sector to address mapping must also rebuild the corresponding key positions. A key position consists of a KSA erase block number and the offset inside the erase block. KSA erase blocks can be logically referenced for wear levelling by storing metadata in the final page of each KSA erase block. This page is written immediately after successfully writing the KSA block and stores the following information: the logical KSA number so that key position references remain valid after updating, and a deletion epoch number so that the most recent version of the KSA block is known.

Generating a correct key-state map when mounting is tied to the internal logic of the FTL. Assuming that the map of logical to physical addresses along with the key positions is correctly created, then it is trivial to iterate over the entries to mark the corresponding keys as assigned. The unmarked positions are then updated to contain new data. The FTL must also generate cryptographically secure random data or be able to receive it from the host. Finally, the file system mounted on the FTL must issue TRIM commands [71] when a sector is deleted, as only the file system has the semantic context to know when a sector is deleted.

12.4 Concluding Remarks

This book provided a thorough examination of the problem of secure data deletion. We focused on two main environments—mobile storage and remote storage—both of which lack secure data deletion despite storing sensitive data and both of which have become extremely relevant in recent years. Our examination of related work not only provides a survey of the literature but further builds a framework useful for comparing secure-deletion solutions and determining their salient features. We then presented a system and adversarial model in which we design and analyze our contributions.

For flash memory, we showed that it is possible for a user to delete data without any modifications to their operating system. When able to modify the operating system, however, DNEFS can be integrated into flash file systems to create a file system with comprehensive and efficient secure data deletion. Our design and implementation of UBIFSec validated DNEFS by showing that it is indeed efficient and unobtrusive to use.

For remote storage, we designed a key disclosure graph: a tool for modelling and reasoning about the adversary's growth of knowledge when storing wrapped encryption keys on a persistent storage medium. We defined useful conditions on mutations for this graph that correspond to ways of updating data structures to easily effect secure deletion. We further designed and built a B-Tree-based solution to analyze the performance in practice. Our final contribution was the analysis of an unreliable securely deleting storage medium. We allowed it to fail in storing and deleting data, and made it robust against these failures by distributing it among multiples nodes. We examined the secure-deletion complications that arose from this and implemented our distributed system for analysis.

Throughout, we provided new ways for thinking about the problem of secure deletion: caveats of focusing on the file as the natural deletion unit, asymmetries in erase/write granularities, mangrove-shaped key disclosure graphs, multiple-access adversaries, keystores with pre-written keys and the consequent existential latency. We hope that these concepts are found useful in the design of future systems.

References

1. "Macy's parade: 'Shredded police papers in confetti'," BBC News. Retrieved from http://www.bbc.co.uk, November 25, 2012.
2. R. Perlman, "The Ephemerizer: Making Data Disappear," Sun Microsystems, Tech. Rep. SMLI TR-2005-140, 2005.
3. Y. Tang, P. Lee, J. Lui, and R. Perlman, "Secure Overlay Cloud Storage with Access Control and Assured Deletion," *Dependable and Secure Computing*, pp. 903–916, 2012.
4. A. Rahumed, H. C. H. Chen, Y. Tang, P. P. C. Lee, and J. C. S. Lui, "A Secure Cloud Backup System with Assured Deletion and Version Control," in *ICPP Workshops*, 2011, pp. 160–167.
5. D. Boneh and R. J. Lipton, "A Revocable Backup System," in *Proceedings of the USENIX Security Symposium*, 1996, pp. 91–96.
6. S. M. Diesburg and A. Wang, "A survey of confidential data storage and deletion methods," *ACM Computing Surveys*, vol. 43, no. 1, pp. 1–37, 2010.
7. S. Swanson and M. Wei, "SAFE: Fast, Verifiable Sanitization for SSDs," University of California, San Diego, Tech. Rep., October 2010.
8. N. Provos, "Encrypting Virtual Memory," in *Proceedings of the 10th USENIX Security Symposium*, 2000, pp. 35–44.
9. R. Perlman, "File System Design with Assured Delete," in *Proceedings of the Network and Distributed System Security Symposium*, 2007.
10. C. Pöpper, D. Basin, S. Capkun, and C. Cremers, "Keeping Data Secret Under Full Compromise Using Porter Devices," in *Proceedings of the 26th Annual Computer Security Applications Conference*, 2010, pp. 241–250.
11. R. Kissel, M. Scholl, S. Skolochenko, and X. Li, "Guidelines for Media Sanitization," September 2006, National Institute of Standards and Technology.
12. D. Perito and G. Tsudik, "Secure Code Update for Embedded Devices via Proofs of Secure Erasure," in *Proceedings of the 15th European Symposium on Research in Computer Security*, D. Gritzalis, B. Preneel, and M. Theoharidou, Eds. Springer, 2010, pp. 643–662.
13. R. Geambasu, T. Kohno, A. A. Levy, and H. M. Levy, "Vanish: increasing data privacy with self-destructing data," in *Proceedings of the 18th USENIX Security Symposium*, 2009, pp. 299–316.
14. R. Geambasu, T. Kohno, A. Krishnamurthy, A. Levy, H. Levy, P. Gardner, and V. Moscaritolo, "New directions for self-destructing data systems," http://systems.cs.columbia.edu/files/wpid-tr10geambasu.pdf, University of Washington, Tech. Rep., 2010.

J. Reardon, *Secure Data Deletion*, Information Security and Cryptography,
DOI 10.1007/978-3-319-28778-2

15. S. Diesburg, C. Meyers, M. Stanovich, M. Mitchell, J. Marshall, J. Gould, A. Wang, and G. Kuenning, "TrueErase: Per-File Secure Deletion for the Storage Data Path," in *Proceedings of the 28th Annual Computer Security Applications Conference*, 2012, pp. 439–448.
16. S. Bauer and N. B. Priyantha, "Secure Data Deletion for Linux File Systems," in *Proceedings of the 10th USENIX Security Symposium*, 2001, pp. 153–164.
17. Mozilla Foundation, "Firefox." [Online]. Available: https://www.mozilla.org/
18. Google, Inc., "Google Chrome." [Online]. Available: http://www.google.com/chrome/
19. Apple, Inc., "Safari." [Online]. Available: http://www.apple.com/safari/
20. European Commission, "Progress on EU Data Protection Reform Now Irreversible Following European Parliament Vote," March 2014. [Online]. Available: http://europa.eu/rapid/press-release_MEMO-14-186_en.htm
21. M. Feldman, "UK Orders Google to Delete Last of Street View Wi-Fi Data," *IEEE Spectrum*, June 24, 2013.
22. A. Herzberg, S. Jarecki, H. Krawczyk, and M. Yung, "Proactive Secret Sharing or How to Cope with Perpetual Leakage," *Advances in Cryptology: CRYPTO*, pp. 339–352, 1995.
23. J. M. Rosenbaum, "In defense of the DELETE key," *The Green Bag*, 2000.
24. A. Greenberg, *This Machine Kills Secrets*. Penguin Books, 2012.
25. S. Garfinkel and A. Shelat, "Remembrance of Data Passed: A Study of Disk Sanitization Practices," *IEEE Security & Privacy*, pp. 17–27, January 2003.
26. M. Wei, L. M. Grupp, F. M. Spada, and S. Swanson, "Reliably Erasing Data from Flash-Based Solid State Drives," in *Proceedings of the 9th USENIX Conference on File and Storage Technologies*, Berkeley, CA, USA, 2011, pp. 105–117.
27. P. T. McLean, "AT Attachment with Packet Interface Extension (ATA/ATAPI-4)," 1998.
28. G. Hughes, T. Coughlin, and D. Commins, "Disposal of disk and tape data by secure sanitization," *IEEE Security & Privacy*, vol. 7, no. 4, pp. 29–34, 2009.
29. T. Gleixner, F. Haverkamp, and A. Bityutskiy, "UBI - Unsorted Block Images," 2006. [Online]. Available: http://www.linux-mtd.infradead.org/doc/ubidesign/ubidesign.pdf
30. S. Loosemore, R. M. Stallman, R. McGrath, A. Oram, and U. Drepper, "The GNU C Library Reference Manual," 2012.
31. Intel Corporation, "Intel Solid-State Drive Optimizer," 2009. [Online]. Available: http://download.intel.com/design/flash/nand/mainstream/Intel_SSD_Optimizer_White_Paper.pdf
32. P. Gutmann, "Secure Deletion of Data from Magnetic and Solid-State Memory," in *Proceedings of the 6th USENIX Security Symposium*, 1996, pp. 77–89.
33. C. Wright, D. Kleiman, and R. S. S. Sundhar, "Overwriting Hard Drive Data: The Great Wiping Controversy," in *Proceedings of the 4th International Conference on Information Systems Security*, 2008, pp. 243–257.
34. Y. Gardi, "MMC card: IOCTL support for Sanitize feature of eMMC v4.5," 2011. [Online]. Available: https://lkml.org/lkml/2011/12/20/219
35. D. Jagdmann, "srm - Linux man page."
36. B. Durak, "wipe(1) - Linux man page."
37. Apple, Inc., "Mac OS X: About Disk Utility's erase free space feature," 2012. [Online]. Available: https://support.apple.com/kb/HT3680
38. J. Garlick, "scrub(1) - Linux man page."
39. S. Garfinkel and D. Malan, "One Big File Is Not Enough: A Critical Evaluation of the Dominant Free-Space Sanitization Technique." in *Privacy Enhancing Technologies, 6th International Workshop*, 2006, pp. 135–151.
40. Oracle Corporation, "About MySQL," 2012. [Online]. Available: http://www.mysql.com/about/
41. Hipp, Wyrick & Company, Inc., "About SQLite," 2012. [Online]. Available: http://www.sqlite.org/about.html
42. P. Stahlberg, G. Miklau, and B. N. Levine, "Threats to privacy in the forensic analysis of database systems," in *Proceedings of the 2007 ACM SIGMOD International Conference on Management of Data*, 2007, pp. 91–102.

43. SQLite, "Pragma statements."
44. R. Card, T. Ts'o, and S. Tweedie, "Design and Implementation of the Second Extended Filesystem," in *Proceedings of the 1st Dutch International Symposium on Linux*. Laboratoire MASI — Institut Blaise Pascal and Massachusetts Institute of Technology and University of Edinburgh, 1995.
45. S. C. Tweedie, "Journaling the Linux ext2fs Filesystem," in *Proceedings of the 4th Annual Linux Expo*, 1998.
46. N. Joukov, H. Papaxenopoulos, and E. Zadok, "Secure Deletion Myths, Issues, and Solutions," in *Proceedings of the 2nd ACM Workshop on Storage Security and Survivability*, 2006, pp. 61–66.
47. E. Zadok and J. Nieh, "FiST: A Language for Stackable File Systems," in *USENIX Technical Conference*, 2000, pp. 55–70.
48. C. Plumb, "shred(1) - Linux man page."
49. N. Joukov and E. Zadokstony, "Adding Secure Deletion to Your Favorite File System," in *Proceedings of the 3rd International IEEE Security In Storage Workshop*, 2005, pp. 63–70.
50. Z. Peterson and R. Burns, "Ext3cow: A Time-Shifting File System for Regulatory Compliance," *ACM Transactions on Storage*, vol. 1, no. 2, pp. 190–212, 2005.
51. Z. Peterson, R. Burns, and J. Herring, "Secure Deletion for a Versioning File System," in *Proceedings of the 4th USENIX Conference on File and Storage Technologies*, 2005.
52. R. L. Rivest, "All-Or-Nothing Encryption and The Package Transform," in *Proceedings of the 4th International Workshop on Fast Software Encryption*, 1997, pp. 210–218.
53. J. A. Halderman, S. D. Schoen, N. Heninger, W. Clarkson, W. Paul, J. A. Calandrino, A. J. Feldman, J. Appelbaum, and E. W. Felten, "Lest we remember: cold-boot attacks on encryption keys," *Communications of the ACM*, vol. 52, pp. 91–98, May 2009.
54. D. Y. Zhu, J. Jung, D. Song, T. Kohno, and D. Wetherall, "TaintEraser: protecting sensitive data leaks using application-level taint tracking," *SIGOPS Oper. Syst. Rev.*, vol. 45, no. 1, pp. 142–154, 2011.
55. A. Whitten and J. D. Tygar, "Why Johnny can't encrypt: a usability evaluation of PGP 5.0," in *Proceedings of the 8th USENIX Security Symposium*, 1999, pp. 169–184.
56. B. Klimt and Y. Yang, "Introducing the Enron Corpus," in *Conference on Email and Anti-Spam*, 2004.
57. V. Hauser, "sfill(1) - Linux man page."
58. M. Mesnier, G. R. Ganger, and E. Riedel, "Object-Based Storage," *Communications Magazine, IEEE*, vol. 41, no. 8, pp. 84–90, 2003.
59. D. Nagle, M. Factor, S. Iren, D. Naor, E. Riedel, O. Rodeh, and J. Satran, "The ANSI T10 object-based storage standard and current implementations," *IBM Journal of Research and Development*, vol. 52, no. 4.5, pp. 401–411, July 2008.
60. P. Maymounkov and D. Mazières, "Kademlia: A Peer-to-Peer Information System Based on the XOR Metric," in *Revised Papers from the First International Workshop on Peer-to-Peer Systems*. Springer, 2002, pp. 53–65.
61. Jedec Solid State Technology Association, "Embedded Multi-Media Card (eMMC) Electrical Standard (5.1)," February 2015. [Online]. Available: http://www.jedec.org/standards-documents/results/jesd84-b51
62. M. Alfeld, W. De Nolf, S. Cagno, K. Appel, D. P. Siddons, A. Kuczewski, K. Janssens, J. Dik, K. Trentelman, M. Walton, and A. Sartorius, "Revealing hidden paint layers in oil paintings by means of scanning macro-XRF: a mock-up study based on Rembrandt's 'An old man in military costume'," *Journal of Analytical Atomic Spectrometry*, vol. 28, pp. 40–51, 2013.
63. A. Ban, "Flash file system," US Patent, no. 5404485, 1995.
64. E. Gal and S. Toledo, "Algorithms and Data Structures for Flash Memories," *ACM Computing Surveys*, vol. 37, pp. 138–163, 2005.
65. Google, Inc., "Google Nexus Phone."
66. Micron Technology, Inc., "Design and Use Considerations for NAND Flash Memory Introduction," Tech. Rep. TN-29-17, 2006. [Online]. Available: https://www.micron.com/resource-details/662bbb05-fa48-4761-b6c9-e3df7750ac93

67. "Memory Technology Devices (MTD): Subsystem for Linux," 2008.
68. W. F. Heybruck, "An Introduction to FAT 16/FAT 32 File Systems," 2007.
69. C. Manning, "How YAFFS Works," 2010.
70. M. Rosenblum and J. K. Ousterhout, "The Design and Implementation of a Log-Structured File System," *ACM Transactions on Computer Systems*, vol. 10, pp. 1–15, 1992.
71. Intel Corporation, "Understanding the Flash Translation Layer (FTL) Specification," 1998.
72. T.-S. Chung, D.-J. Park, S. Park, D.-H. Lee, S.-W. Lee, and H.-J. Song, "A survey of Flash Translation Layer," *Journal of Systems Architecture*, pp. 332–343, 2009.
73. G. Goodson and R. Iyer, "Design Tradeoffs in a Flash Translation Layer," in *Proceedings of the Workshop on the Use of Emerging Storage and Memory Technologies*, 2010.
74. D. Woodhouse, "JFFS: The Journalling Flash File System," in *Proceedings of the Ottawa Linux Symposium*, 2001. [Online]. Available: http://sources.redhat.com/jffs2/jffs2.pdf
75. A. Hunter, "A Brief Introduction to the Design of UBIFS," 2008.
76. C. Lee, D. Sim, J.-Y. Hwang, and S. Cho, "F2FS: A New File System for Flash Storage," in *Proceedings of the 13th USENIX Conference on File and Storage Technologies*, 2015, pp. 273–286.
77. J. Lee, S. Yi, J. Heo, and H. Park, "An Efficient Secure Deletion Scheme for Flash File Systems," *Journal of Information Science and Engineering*, pp. 27–38, 2010.
78. Open NAND Flash Interface, "Open NAND Flash Interface Specification, version 3.0," 2011. [Online]. Available: http://onfi.org/specifications/
79. L. M. Grupp, A. M. Caulfield, J. Coburn, S. Swanson, E. Yaakobi, P. H. Siegel, and J. K. Wolf, "Characterizing flash memory: anomalies, observations, and applications," in *IEEE/ACM International Symposium on Microarchitecture*, New York, NY, USA, 2009, pp. 24–33.
80. Samsung, Inc., "Samsung Galaxy S Phone."
81. R. Stoica, M. Athanassoulis, R. Johnson, and A. Ailamaki, "Evaluating and Repairing Write Performance on Flash Devices," in *Proceedings of the 5th International Workshop on Data Management on New Hardware*, 2009, pp. 9–14.
82. S. Boboila and P. Desnoyers, "Write Endurance in Flash Drives: Measurements and Analysis," in *Proceedings of the 8th USENIX Conference on File and Storage Technologies*, 2010, pp. 115–128.
83. R. Entner, "International Comparisons: The Handset Replacement Cycle," Recon Analytics, Tech. Rep., 2011.
84. A. W. Appel, "Simple generational garbage collection and fast allocation," *Software Practice and Experience*, vol. 19, pp. 171–183, 1989.
85. C. Fruhwirth, "New methods in hard disk encryption," Vienna University of Technology, Tech. Rep., 2005.
86. A. J. Menezes, P. C. van Oorschot, and S. A. Vanstone, *Handbook of Applied Cryptography, Chapter 7: Block Ciphers*. CRC Press, 1997.
87. T. Cormen, C. Leiserson, and R. Rivest, *Introduction to Algorithms*. MIT Press, 1998.
88. J. Reardon, "UBIFSec Implementation," 2011. [Online]. Available: http://www.syssec.ethz.ch/content/dam/ethz/special-interest/infk/inst-infsec/system-security-group-dam/research/secure_deletion/ubifsec.patch
89. E.M. Hoover, "The Measurement of Industrial Localization," *The Review of Economics and Statistics*, pp. 162–171, 1936.
90. J. Chow, B. Pfaff, T. Garfinkel, and M. Rosenblum, "Shredding Your Garbage: Reducing Data Lifetime Through Secure Deallocation," in *Proceedings of the 14th USENIX Security Symposium*, 2005.
91. J. Clark and A. Essex, "CommitCoin: Carbon Dating Commitments with Bitcoin," in *International Conference on Financial Cryptography and Data Security*. Springer, 2012, pp. 390–398.
92. G. D. Crescenzo, N. Ferguson, R. Impagliazzo, and M. Jakobsson, "How to Forget a Secret," in *Proceedings of the 16th Annual Symposium on Theoretical Aspects of Computer Science*, ser. Lecture Notes in Computer Science. Springer, 1999, pp. 500–509.

93. C. Cachin, K. Haralambiev, H. Hsiao, and A. Sorniotti, "Policy-based secure deletion," in *Proceedings of the 2013 ACM SIGSAC Conference on Computer & Communications Security*, 2013, pp. 259–270.
94. A. Shamir, "How to share a secret," *Communications of the ACM*, vol. 22, no. 11, pp. 612–613, 1979.
95. R. Geambasu, A. A. Levy, T. Kohno, A. Krishnamurthy, and H. M. Levy, "Comet: An Active Distributed Key-Value Store," in *Proceedings of the 9th USENIX conference on Operating Systems Design and Implementation*, 2010, pp. 323–336.
96. S. Wolchok, O. S. Hoffman, N. Henninger, E. W. Felten, J. A. Haldermann, C. J. Rossback, B. Waters, and E. Witchel, "Defeating Vanish with Low-Cost Sybil Attacks Against Large DHTs," in *Proc. 17th Network and Distributed System Security Symposium*, Feb. 2010.
97. W. T. Tutte, *Graph Theory*, ser. Encyclopedia of Mathematics and Its Applications. Addison-Wesley Publishing Company, 1984.
98. D. Comer, "The Ubiquitous B-Tree," *ACM Computing Surveys*, vol. 11, pp. 121–137, 1979.
99. R. Fagin, J. Nievergelt, N. Pippenger, and H. R. Strong, "Extendible hashing—a fast access method for dynamic files," *ACM Transactions on Database Systems*, vol. 4, no. 3, pp. 315–344, 1979.
100. O. Rodeh, "B-trees, shadowing, and clones," *Trans. Storage*, vol. 3, no. 4, pp. 2:1–2:27, 2008.
101. R. McDougall and J. Mauro, "FileBench," 2005. [Online]. Available: www.solarisinternals.com/si/tools/filebench/
102. L. A. Bélády, "A study of replacement algorithms for virtual-storage computer," *IBM Systems Journal*, vol. 5, no. 2, pp. 78–101, 1966.
103. R. C. Merkle, "A certified digital signature," in *Proceedings on Advances in Cryptology*, ser. CRYPTO '89. Springer, 1989, pp. 218–238.
104. E. Mykletun, M. Narasimha, and G. Tsudik, "Authentication and Integrity in Outsourced Databases," *Trans. Storage*, pp. 107–138, 2006.
105. Z. Wilcox-O'Hearn and B. Warner, "Tahoe: The Least-Authority Filesystem," in *Proceedings of the 4th ACM International Workshop on Storage Security and Survivability*, 2008, pp. 21–26.
106. FUSE Developers, "The Filesystem in Userspace Website," 2014. [Online]. Available: http://fuse.sourceforge.net
107. T. Dierks and E. Rescorla, "The Transport Layer Security (TLS) Protocol Version 1.2," Internet Requests for Comments, RFC 5246, May 2000.
108. W. Diffie and M. Hellman, "New directions in cryptography," *IEEE Transactions on Information Theory*, vol. 22, no. 6, pp. 644–654, 1976.
109. D. Cooper, S. Santesson, S. Farrell, S. Boeyen, R. Housley, and T. Polk, "Internet X.509 Public Key Infrastructure Certificate and Certificate Revocation List (CRL) Profile," Internet Requests for Comments, RFC 6818, May 2008.
110. Amazon Inc., "Amazon Webservices," 2014. [Online]. Available: https://aws.amazon.com
111. Dropbox, Inc., "The Dropbox Website," 2014. [Online]. Available: https://www.dropbox.com
112. Google, Inc., "The Google Drive Website," 2014. [Online]. Available: https://drive.google.com
113. N. A. Lynch, *Distributed Algorithms*. Morgan Kaufmann, 1996.
114. L. Lamport, R. Shostak, and M. Pease, "The Byzantine Generals Problem," *ACM Transactions on Programming Languages and Systems*, vol. 4, no. 3, pp. 382–401, 1982.
115. B. Chor, S. Goldwasser, S. Micali, and B. Awerbuch, "Verifiable Secret Sharing and Achieving Simultaneity in the Presence of Faults," in *Proceedings of the 26th Annual Symposium on Foundations of Computer Science*, 1985, pp. 383–395.
116. P. Feldman, "A practical scheme for non-interactive verifiable secret sharing," in *Proceedings of the 28th Annual Symposium on Foundations of Computer Science*, 1987, pp. 427–438.
117. M. Oberhumer, "The LZO Website," 2014. [Online]. Available: http://www.oberhumer.com/opensource/lzo/

118. A. J. Menezes, P. C. van Oorschot, and S. A. Vanstone, *Handbook of Applied Cryptography, Chapter 5: Pseudorandom Bits and Sequences*. CRC Press, 1997.
119. V. Shoup, "NTL: A Library for Doing Number Theory," 2015. [Online]. Available: http://www.shoup.net/ntl/
120. INRIA, "gf2x," 2012. [Online]. Available: http://gf2x.gforge.inria.fr/
121. xmlrpc-c Developers, "The XML-RPC for C and C++ Website," 2014. [Online]. Available: http://xmlrpc-c.sourceforge.net
122. stunnel Developers, "The stunnel Website," 2014. [Online]. Available: http://www.stunnel.org
123. E. Rescorla, "HTTP Over TLS," Internet Requests for Comments, RFC 2818, May 2000.
124. openssl Developers, "The OpenSSL Website," 2014. [Online]. Available: www.openssl.org
125. H. Ritzdorf, N. Karapanos, and S. Capkun, "Assisted Deletion of Related Content," in *Proceedings of the 30th Annual Computer Security Applications Conference*. ACM, 2014, pp. 206–215.
126. G. G. Richard III and V. Roussev, "Scalpel: A frugal, high performance file carver," in *Digital Forensic Research Workshop*, 2005.
127. S. Garfinkel, "Carving contiguous and fragmented files with fast object validation," *Digital Investigation*, vol. 4, Supplement, pp. 2–12, 2007.
128. A. Pal and N. Memon, "The Evolution of File Carving," *IEEE Signal Processing Magazine*, vol. 26, no. 2, pp. 59–71, 2009.
129. K. Shanmugasundaram and N. Memon, "Automatic Reassembly of Document Fragments via Context Based Statistical Models," in *Proceedings of the 19th Annual Computer Security Applications Conference*. ACM, 2003, pp. 152–159.
130. A. Pal, K. Shanmugasundaram, and N. Memon, "Automated reassembly of fragmented images," in *Proceedings of the 2003 International Conference on Multimedia and Expo*. IEEE Computer Society, 2003, pp. 625–628.
131. A. D. McDonald and M. G. Kuhn, "StegFS: A Steganographic File System for Linux," in *Proceedings of the 3rd International Workshop on Information Hiding*, ser. Lecture Notes in Computer Science. Springer, 1999, pp. 462–477.
132. A. Skillen and M. Mannan, "Mobiflage: Deniable Storage Encryption for Mobile Devices," *IEEE Transaction on Dependable and Secure Computing*, vol. 11, no. 3, pp. 224–237, May 2014.
133. T. Peters, M. Gondree, and Z. N. J. Peterson, "DEFY: A Deniable, Encrypted File System for Log-Structured Storage," in *22nd Annual Network and Distributed System Security Symposium*, 2015.
134. S. Bajaj and R. Sion, "HIFS: History Independence for File Systems," in *Proceedings of the 20th ACM SIGSAC Conference on Computer & Communications Security*, 2013, pp. 1285–1296.
135. N. Karvelas and A. Kiayias, "Efficient Proofs of Secure Erasure," in *9th International Conference on Security and Cryptography for Networks*. Springer, 2014, pp. 520–537.
136. F. Hao, D. Clarke, and A. Zorzo, "Deleting Secret Data with Public Verifiability," *IEEE Transactions on Dependable and Secure Computing*, 2015.

Glossary

address (storage medium) A location where data is stored. These can be logical, where the address is simply a handle by which data can be later recalled, or physical, where the address describes an actual position on a material where data is stored.

adversary-controlled access time See *unpredictable access time*.

all-or-nothing transformation A cryptographic technique that ensures that if any part of the cipher text is missing, then the entire message cannot be decrypted even with the decryption key.

analog remnants See *remnants*.

ancestor (graph theory) A relation between two vertices in a directed graph; a vertex is an ancestor of another if there exists a directed path from the ancestor vertex to the other vertex. All vertices are ancestors of themselves.

arborescence (graph theory) In graph theory, a directed graph with two properties: (i) its underlying undirected structure forms a tree (i.e., has no cycles), and (ii) either all edges diverge or converge towards a root node.

assigned (keystore) A key position state corresponding to a key value that has been provided to a client to encrypt data.

attack surface The adversary's interface to the storage medium when it is attempting to recover deleted data.

B-Tree A self-balancing search tree data structure widely used in storage settings.

bad (erase) block In flash memory, an erase block that is no longer able to store data because it has been worn out with erasures.

batching The act of performing a sequence of operations in rapid succession.

behavioural properties Properties of secure-deletion solutions. These include deletion granularity, deletion scope, the effect on the storage medium's lifetime,

J. Reardon, *Secure Data Deletion*, Information Security and Cryptography,
DOI 10.1007/978-3-319-28778-2

deletion latency, and efficiency. Secure deletion solutions can be compared based on their behavioural properties.

Byzantine failure A failure where an entity behaves in an arbitrary way, e.g., by issuing spurious well-formed requests.

clocked keystore A storage medium whose purpose is to assign and securely delete cryptographic keys, such that keys that are marked for deletion are collected and securely deleted in batch during a period clock operation.

coercive adversary A secure deletion adversary that compromises not only the user's storage media but also any keys or passphrases that may grant access; securely deleting data against such an adversary is equivalent to securely deleting the data such that the users themselves cannot recover it.

compaction The act of reclaiming wasted space on a unit of memory, for instance, by taking units of memory sparsely filled with data and copying the data so it takes up less space on a densely filled unit of memory.

complementary key pools (keystore) A pair of key pools such that no keystore machine stores data belonging to both pools.

computational bounded (adversary) A secure deletion adversary with limited time and resources. For our purposes, it means that the adversary cannot decrypt data without knowledge of the corresponding encryption key or break other cryptographic primitives.

computational unbounded (adversary) A secure deletion adversary with unlimited time and resources. For our purposes, it means that securely deleting an encryption key does not imply the secure deletion of the corresponding data.

content store A large persistent storage medium used to store bulk data.

controller layer (storage) The hardware that interacts directly with the physical material that stores data.

crash The sudden cessation of computation without the opportunity to save state, e.g., due to a power failure.

creation (data item) The beginning of a data item's valid lifetime. The time at which it should be now stored on a storage medium.

cryptographic digest A characteristic fixed-size binary string for an arbitrarily sized binary string; a cryptographic digest function computes a short collision-resistant *hash* of a string and it is computationally infeasible to reverse this function or find two strings with the same digests.

cycle (graph theory) A *non-degenerate walk* in a *graph* that begins and ends at the same *vertex*.

data item An atomic unit of data. A data item is created, read, and discarded in its entirety. Updates imply discarding the old item and creating a new one.

data lifetime The period of time between a data item's *creation* and its subsequent *discard*.

data node Alternative name for *data item*.

data node encrypted file system (DNEFS) A file system we propose that encrypts each unit of data with its own unique encryption key.

delete An ambiguous term often conveying the notion of *secure delete* while implemented as *discard*.

deletion epoch A period of time between executions of a secure-deletion procedure. Solutions with deletion epochs collect data to be deleted during the epoch and securely delete them simultaneously.

descendant (graph theory) A relation between two vertices in a directed graph; a vertex is a descendant of another if there exists a directed path from the other vertex to the descendant. All vertices are descendants of themselves.

device driver (storage) A software abstraction that consolidates access to different types of hardware by exposing a single common interface.

device lifetime (solution) The physical wear or damage caused on a storage medium based on executing a secure-deletion solution. Shredding a hard drive, for example, causes *complete wear*, while overwriting it causes nearly no wear.

directed acyclic graph A graph whose edges are directed and which contains no cycles.

discard The action of indicating something should be deleted.

discard (data item) The end of a data item's valid lifetime.

discarded (keystore) A key position state corresponding to a key value that should be securely deleted at the next opportunity.

efficiency (solution) A polymorphic behavioural property of secure-deletion solutions. This is any cost metric, such as the solution's execution time, battery consumption, etc.

environmental assumption (solution) Assumptions made by a secure-deletion solution that, if held, ensure the correct functioning of the system that comes with additional behavioural properties. If the assumptions do not hold, then no guarantee exists. Assumptions are the adversary's maximal characteristics and the interface offered by the storage medium's access layer.

erase block A division of flash memory that forms the unit of erasure. 256 KiB is a representative size of an erase block, though different memories have different sizes. Erase blocks can become *bad* if too many erasures are performed on them. Erase blocks themselves are divided into *pages*.

erase block reallocation The act of taking an erase block that previously stored data and erasing it so that it may store new data.

existential latency (solution) The time before a data item's creation when it can be compromised by an adversarial attack. This is manifested by the adversary learning keys used to encrypt future data, and then later learning the encrypted data itself.

file system (storage) A system for organizing data into files, which are in turn stored in a hierarchy of directories.

flash file system (FFS) A file system designed specifically for the nuances of flash memory in the file system. Flash file systems are used instead of a *flash translation layer*.

flash translation layer (FTL) A logical remapping layer for flash memory that handles the fact that flash cannot overwrite data but it must be instead erased at a high granularity. FTLs are usually implemented in hardware and are widely used in consumer flash devices.

forward secrecy The desirable property that ensures that the compromise of a user's long-term cryptographic key does not affect the confidentiality of past communications. This is often achieved by protecting the communications with session keys authentically negotiated using the long-term key.

fraudulent failure (keystore) A failure where a keystore node invents a state change (i.e., assign or discard) that did not actually occur.

full scope (solution) A secure-deletion solution that securely deletes all unused blocks on a storage medium.

garbage collection See *compaction*

granularity (solution) The precision by which secure deletion can be achieved, e.g., individual data items or an entire storage medium.

graph A set of vertices and a set of pairs of distinct vertices called edges.

handle (data) A reference to some data that is used to access it.

hash See *cryptographic digest*.

immediate (solution) A secure-deletion solution which securely deletes data without any significant deletion latency.

in-place update A storage medium whose interface allows stored data to be directly overwritten: its logical address space is equal to its physical storage locations.

indegree (graph theory) The number of incoming edges that a vertex has.

input/output control (ioctl) An operating system function that permits low-level device-specific operations that are not available in a standard interface.

interface The set of things one can do when interacting with an object; e.g., buttons and dials on a tactile interface.

invalid (data item) A data item that has not yet been created or has been previously discarded; a data item that should not be stored on any storage medium.

journalling file system A file system that stores all recent changes in a well-known place so that, in the event of a crash, the correct file system can be recreated without having to examine the entire contents of the file system.

key class A sequence of key pools that define the composition of a key recipe; each key recipe has a single key class.

key pool A grouping of key values; each key value has a single key pool. All attributes by which key values can be discriminated are encapsulated by its key pool.

key position (keystore) A reference to a key value, associated with a state that is either *unused*, *assigned*, or *discarded*.

key recipe A sequence of key positions that correspond to the key values that must be retrieved in order to create the encryption key.

key value (keystore) A random binary string suitable for use as an encryption key.

key wrapping The technique of encrypting an encryption key with another key; the latter key is said to wrap the former.

keystore A storage medium whose purpose is to assign and securely delete cryptographic keys.

keystore node (keystore) An entity in the distributed system that makes up the keystore. Each node is responsible for storing a subset of key values, which are distributed for robustness.

laboratory attacks (NIST) Hardware attacks to recover data from storage media: the adversary can modify hardware that controls the storage media and use any forensic equipment at its disposal.

log-structured file system A type of file system in which all data and metadata is stored as a sequence of changes from an initial empty state.

mangrove (graph theory) A directed graph such that the subgraph induced by each vertex forms an arborescence directed away from it.

memory technology device (MTD) The term used to refer to flash memory devices in Linux.

multimedia card (MMC) The generic name for flash-memory-based removable storage cards, e.g., SD cards.

multiple access (adversaries) A secure-deletion adversary that may compromise the storage medium at multiple points in time.

negligent failure (keystore) A failure where a keystore node neglects to report a state change (i.e., assign or discard) that occurred.

NIST guidelines (for media sanitization) A detailed standard for securely deleting data stored on a wide variety of storage media.

non-coercive adversary A secure-deletion adversary that only obtains a user's storage media but not their secret keys or passphrases; encryption is generally sufficient against such an adversary.

non-degenerate walk (graph theory) A non-empty sequence of edges in a graph such that removing the last edge forms a walk and the last edge begins at the terminus of the walk formed by removing the edge.

not valid (data item) See *invalid*.

object store device (OSD) A storage system that organizes data as objects that include their own metadata.

outdegree (graph theory) The number of outgoing edges that a vertex has.

path (graph theory) A *walk* that does not visit the same vertex twice.

per-block level (solution) A secure-deletion solution that can securely delete the atomic unit of data, e.g., a file system block.

per-file level (solution) A secure-deletion solution that can securely delete only entire files.

per-storage-medium level (solution) A secure-deletion solution that can securely delete only the entire storage medium.

periodic (solution) A secure deletion solution that executes intermittently and provides a *deletion latency* based on the time period between subsequent executions. Discarded data is left in place. When the periodic solution executes, the data is securely deleted in *batch*. Periodically running a secure-deletion solution punctuates time into *deletion epochs*.

persistent storage medium A storage medium that is unable to delete any data that is stored on it. Accessing such a storage medium reveals its entire history of write operations.

physical destruction A secure-deletion technique that ensures data is irrecoverable by having the storage medium cease to usefully exist, e.g., incineration, pulverization, etc.

physical layer (storage) The actual material that stores data, e.g., paper, tape.

portable operating system interface (POSIX) An abstract operating system and core set of operating system features used for portability across implementations.

predictable access time (adversary) An adversary that gains access to a user's storage media at a predictable time, thereby allowing the execution of an extraordinary secure-deletion procedure.

random access Accessing stored data such that the next address read is independent of the one being currently read; with analogy to a book: reading pages randomly.

reachable (graph theory) A relation on two vertices in a directed graph which is true if there is a path from the former vertex to the latter.

remnants Data that is left behind after a deletion operation, such as indentations on a page or streaks on a blackboard.

replication set (keystore) The set of keystore nodes that are responsible for storing a key value for a particular key position.

robust-keyboard attacks (NIST) Software attacks on a storage medium; it remains attached to its controller and can only be interacted with at the device driver layer.

secure deletion The act of removing data stored on a storage medium such that the data is irrecoverable.

sequential access Accessing stored data such that the next address read follows sequentially from the one being currently read; with analogy to a book: reading pages in order.

single access (adversaries) A secure-deletion adversary that may compromise the storage medium only once.

solid-state drive (SSD) A storage medium technology that uses flash memory to store data, commonly used instead of magnetic spinning-disk technology as it does not incur latency for seek operations.

spinning-disk storage medium A storage medium that stores data on a rotating disk and accesses it with fixed read and write heads. It spins the disk to the appropriate position when reading. Spinning-disk storage is noted for having a considerable performance penalty when performing *random access*.

storage medium Anything that stores data, such as a hard drive, a phone, a brain, or a blackboard.

TRIM commands Notifications from the file system to the device driver to inform the latter about discarded data blocks, e.g., the deletion of a large file. This prevents a thrashing effect in flash memory where deleted data continues to be copied around because the device driver can only infer its deletion when its *logical address* is overwritten.

UBI file system (UBIFS) A file system designed to use the UBI interface for flash memory.

unlinking In file systems, the act of removing a file from a system by simply removing its handle from its containing directory.

unmapped (storage region) A part of a storage medium that is not referenced by any data in the file system.

unordered block images (UBI) An interface for MTD devices that adds a logical remapping of erase blocks to support atomic updates and automatic wear levelling.

unpredictable access time (adversary) An adversary that gains access to a user's storage media at an unpredictable time, thereby preventing any extraordinary secure-deletion procedure.

unused (keystore) A *key position* state corresponding to a fresh random *key value* that has not been provided to any client.

user interface (storage) The interface given to a user by applications that store and delete data on a system, which is manipulated by devices such as keyboards and mice.

user-controlled access time See *predictable access time*.

valid (data item) A data item that has been created and has not yet been discarded; a data item that should be reliably stored.

versioning file system A file system that automatically stores changes to data separately from the original data so that the history of changes is available.

walk (graph theory) Either a *non-degenerate walk* or a zero-length sequence of edges that (technically) begins and ends at the same vertex.

wear levelling The effort to spread the damage caused by erasing flash memory fairly over the entire storage medium.

write-once, read-many (WORM) A type of storage medium that, once written, cannot be changed or deleted.

zero overwriting A frequently used technique in secure deletion where the data's storage location is overwritten with binary zeros.

Index

INPLACE model, 35
PERSISTENT model, 36, 116, 144
SECDEL-CLOCK-EXIST, 135
SECDEL-CLOCK-EXIST model, 35, 144
SECDEL-CLOCK model, 35, 67
SECDEL model, 35, 116
SEMIPERSISTENT model, 36, 64
nandsim, 61, 96

access time, 24
access token (AT), 75, 147
acyclic, 117
add (B-Tree), 130
adversarial model, 23, 36
adversarial resistance, 27
adversary-controlled access time, 24
all-or-nothing transformation, 21, 181
analog remnants, 5, 16
ancestor, 117
arborescence, 117
assign KV, 75
assign quorum, 151
assigned KP, 77, 156
assigner (keystore node), 148
ATA, 12
atomic updates (UBI), 88
attack surface, 23

B-Tree, 129, 130
bad block (flash), 48
ballooning, 65
ballooning–purging hybrid, 67
batching, 54, 83
behavioural properties, 27, 31
Byzantine failure, 153

caching, 99, 137
class (keystore), 159
classes (file), 110
client, 144, 146
clock operation, 77
clocked keystore, 77, 145, 147
cloud storage, 106
coercive adversary, 24
coercive attack, 36, 75
commit and replay, 89, 92
compaction, 49, 53, 58
complementary key pool, 157
complete wear, 29
computational bound, 25
computationally bounded adversary, 25
computationally unbounded adversary, 25
content store, 144, 145
controller layer, 12
creation time, 33
credential revelation, 24
cryptographic digest, 131
cycle, 117

data item, 6, 33
data lifetime, 33, 162
data node encrypted file system, 73
DEFY, 180
deletion epoch, 78
deletion granularity, 27
deletion latency, 29, 37, 62, 70, 77
deniable storage, 180
denial-of-service (DoS), 163
deployment requirements, 31
descendant, 117
device layer, 14
directed acyclic graph, 117
dirty node, 133
discard, 5

J. Reardon, *Secure Data Deletion*, Information Security and Cryptography,
DOI 10.1007/978-3-319-28778-2

discard KV, 75
discard quorum, 151
discard time, 33
discarded KP, 77, 156
distributed clocked keystore, 147
distributed hash table (DHT), 112

edge, 116
efficiency, 30
entropy pool, 164
environmental assumptions, 27, 31
Ephemerizer, 109
erase block, 48
erase block reallocation, 59
erase granularity, 52
evil maid, 24
existential latency, 37, 77, 109, 129, 135
expiration-time granularity, 109
ext2, 20
ext3, 20

F-Table, 110
F2FS, 52
factory reset, 53
fade, 110, 111
file carving, 180
file system layer, 14
filling, 18
flash file system, 49, 52
flash memory, 48
flash translation layer, 49, 51
forward secrecy, 5, 145
fraudulent failure, 153
FTL, 12, 49, 51, 184
full (subdigraph), 117
FUSE, 169
fuse (B-Tree), 130

garbage collection, 49, 58
granularity (flash), 48
Gutmann, 16

handle (data item), 33
hardware DNEFS, 184
hash, 131
hashing, 118
HIFS, 181
history independence, 181

immediate deletion, 29, 35, 53
in-place updates, 20, 49
indegree, 116
induced graph, 117
information deletion, 179
initialization vector, 82
interface, 6
invalid (data item), 33

JFFS, 52

key class, 159
key disclosure graph (KDG), 118
key pool, 157
key recipe, 157
key storage area, 73, 75
key value (KV), 75, 145, 147
key wrapping, 109, 118
key-state map, 79
key-value map, 129, 136
keystore, 75, 145
keystore file system (KSFS), 169
keystore node, 145
keystore properties, 77, 147
KSA, 83
KSA update, 79

laboratory attacks, 23
layers, 12, 14, 27
legislative requirements, 4, 47, 106
lifetime, 29, 33
log-structured file system, 49, 51
logical erase block (LEB), 88
long-term KSA, 90

magnetic tape, 52
mangrove, 117
Merkle tree, 131, 132
metadata, 33
MMC, 12, 51, 53
Mobiflage, 180
modify (B-Tree), 130
MTD, 14, 52, 88
mtdblock, 49
multiple-access adversary, 24
mutated graph, 120
MySQL, 19

name function, 116
negligent failure, 153
network block device (NDB), 136
NIST guidelines, 4, 15
non-coercive adversary, 24
non-coercive attacks, 146
number of accesses, 24

object store device (OSD), 33, 136
origin, 117
outdegree, 116
over-reporting threshold, 155
overwriting, 17

page (flash), 48
path, 117
per-data-item granularity, 27
per-file granularity, 27, 54
per-file-class granularity, 110
per-storage-medium granularity, 27
perfect forward secrecy, 144, 164
periodic deletion, 29, 35, 54
persistent storage medium, 105, 106
physical erase block (PEB), 88
physical layer, 12
plausible deniability, 180
policy-based secure deletion, 112
polynomial interpolation, 161
porter devices, 111
POSIX, 14
power consumption, 71, 98
predictable access time, 24
proactive security, 5
program disturb (flash), 48, 55
programming (flash), 48
proof of secure erasure, 182
provable deletion, 182
pull phase (sync), 150
purgefs, 21
purging, 64
push phase (sync), 150

reachable, 117
read KV, 75
rebalance (B-Tree), 130
recipe, 157
rekeying, 109
remote storage, 106
remove (B-Tree), 130
replication set, 148
replicator (keystore node), 148
revocable backup system, 109
robust-keyboard attacks, 23
root, 117

sanitize, 53
scope, 28
scrub budget, 55
scrubbing, 54
SCSI, 12
secret sharing, 161
secure erase, 16, 53
securely deleting storage medium, 106
security goal, 37
security initialize, 16
sensitive file, 28
shadowed updates, 131
shadowing, 120
shadowing graph mutation, 121
shadowing graph mutation chain, 125
Shamir secret sharing, 161
short-term KSA, 90
single-access adversary, 24
skeleton tree, 133
some wear, 29, 70, 96
split (B-Tree), 130
SQLite, 19
srm, 17
SSD, 12
steganographic storage, 180
storage interfaces, 12, 27
storage layers, 12, 14, 27
storage medium, 6
subdigraph, 117
synchronization protocol, 150
system integration, 27

tainting, 28
targeted scope, 28
terminus, 117
tree node cache, 89
TRIM commands, 22
TrueErase, 22

UBI, 14, 88
UBIFS, 52, 88
UBIFSec, 90
unchanged (wear), 29
unlink (file), 5
unmapped, 51, 53
unpredictable access time, 24
untargeted scope, 28
unused KP, 77
user interface layer, 14
user level, 58, 63
user-controlled access time, 24

valid (data item), 33
Vanish, 112
versioning, 133
versioning file system, 21
vertex, 116
virtual block device, 51, 129
virtual storage device, 137

walk, 117
wear levelling, 48, 88, 96
wipe, 17
WORM medium, 5
write granularity, 52

YAFFS, 52, 58
YAFFS1, 59
YAFFS2, 59

zero overwriting, 17, 54